CAMBRIDGE LIBRARY COLLECTION

Books of enduring scholarly value

Physical Sciences

From ancient times, humans have tried to understand the workings of
the world around them. The roots of modern physical science go back to
the very earliest mechanical devices such as levers and rollers, the mixing
of paints and dyes, and the importance of the heavenly bodies in early
religious observance and navigation. The physical sciences as we know them
today began to emerge as independent academic subjects during the early
modern period, in the work of Newton and other 'natural philosophers',
and numerous sub-disciplines developed during the centuries that followed.
This part of the Cambridge Library Collection is devoted to landmark
publications in this area which will be of interest to historians of science
concerned with individual scientists, particular discoveries, and advances in
scientific method, or with the establishment and development of scientific
institutions around the world.

Telescopic Work for Starlight Evenings

William F. Denning (1848–1931) was a British astronomer famous for his
planetary observations and meteor studies. Elected president of the Liverpool
Astronomical Society in 1887, he wrote a series of articles on telescopes
for the society's journal, which were brought together and republished in
1891 under the title *Telescopic Work for Starlight Evenings*. Intended as a
contribution to popular astronomy, this book provides a varied introduction
to telescopes and their usage. The opening essay traces the development of
the telescope from antiquity, through Galileo and Newton's contributions
in the seventeenth century, to contemporary progress in astronomy. Other
chapters provide practical advice for conducting planetary observation
and detailed studies of particular planets, as well as facts and figures about
meteors and how to compute their orbit. This book provides a fascinating
insight into the evolution of astronomy and will be a valuable resource for
historians of science and amateur astronomers.

Cambridge University Press has long been a pioneer in the reissuing of out-of-print titles from its own backlist, producing digital reprints of books that are still sought after by scholars and students but could not be reprinted economically using traditional technology. The Cambridge Library Collection extends this activity to a wider range of books which are still of importance to researchers and professionals, either for the source material they contain, or as landmarks in the history of their academic discipline.

Drawing from the world-renowned collections in the Cambridge University Library, and guided by the advice of experts in each subject area, Cambridge University Press is using state-of-the-art scanning machines in its own Printing House to capture the content of each book selected for inclusion. The files are processed to give a consistently clear, crisp image, and the books finished to the high quality standard for which the Press is recognised around the world. The latest print-on-demand technology ensures that the books will remain available indefinitely, and that orders for single or multiple copies can quickly be supplied.

The Cambridge Library Collection will bring back to life books of enduring scholarly value (including out-of-copyright works originally issued by other publishers) across a wide range of disciplines in the humanities and social sciences and in science and technology.

Telescopic Work
for Starlight Evenings

WILLIAM F. DENNING

CAMBRIDGE
UNIVERSITY PRESS

CAMBRIDGE UNIVERSITY PRESS

Cambridge, New York, Melbourne, Madrid, Cape Town, Singapore,
São Paolo, Delhi, Dubai, Tokyo

Published in the United States of America by Cambridge University Press, New York

www.cambridge.org
Information on this title: www.cambridge.org/9781108014137

© in this compilation Cambridge University Press 2010

This edition first published 1891
This digitally printed version 2010

ISBN 978-1-108-01413-7 Paperback

STANMORE OBSERVATORY.

TELESCOPIC WORK

FOR

STARLIGHT EVENINGS.

BY

WILLIAM F. DENNING, F.R.A.S.
(FORMERLY PRESIDENT OF THE LIVERPOOL ASTRONOMICAL SOCIETY).

"To ask or search I blame thee not, for heaven
Is as the book of God before thee set,
Wherein to read his wondrous works."
MILTON.

LONDON:

TAYLOR AND FRANCIS, RED LION COURT, FLEET STREET.

1891.

ALERE FLAMMAM.

PREFACE.

————◆————

IT having been suggested by some kind friends that a series of articles on "Telescopes and Telescopic Work," which I wrote for the 'Journal of the Liverpool Astronomical Society' in 1887–8, should be reprinted, I have undertaken the revision and re-arrangement of the papers alluded to. Certain other contributions on "Large and Small Telescopes," "Planetary Observations," and kindred subjects, which I furnished to 'The Observatory' and other scientific serials from time to time, have also been included, and the material so much altered and extended that it may be regarded as virtually new matter. The work has outgrown my original intention, but it proved so engrossing that it was found difficult to ensure greater brevity.

The combination of different papers has possibly had the effect of rendering the book more popular in some parts than in others. This is not altogether unintentional, for the aim has been to make the work intelligible to general readers, while also containing facts and figures useful to amateur astronomers. It is merely intended as a contribution to popular astronomy, and asserts no rivalry with existing works, many of which are essentially different in plan. If any excuse were, however, needed for the issue of this volume it might be found in the rapid progress of astronomy, which requires that new or revised works should be published at short intervals in order to represent existing knowledge.

The methods explained are approximate, and technical points have been avoided with the view to engage the interest of beginners who may find it the stepping-stone to more advanced works and to more precise methods. The object will be realized if observers derive any encouragement from its descriptions or value from its references, and the author sincerely hopes that not a few of his readers will experience the same degree of pleasure in observation as he has done during many years.

No matter how humble the observer, or how paltry the telescope, astronomy is capable of furnishing an endless store of delight to its adherents. Its influences are elevating, and many of its features possess the charms of novelty as well as mystery. Whoever contemplates the heavens with the right spirit reaps both pleasure and profit, and many amateurs find a welcome relaxation to the cares of business in the companionship of their telescopes on "starlight evenings."

The title chosen is not, perhaps, a comprehensive one, but it covers most of the ground, and no apology need be offered for dealing with one or two important objects not strictly within its scope.

For many of the illustrations I must express my indebtedness to the Editors of the 'Observatory,' to the Council of the R.A.S., to the proprietors of 'Nature,' to Messrs. Browning, Calver, Cooke & Sons, Elger, Gore, Horne Thornthwaite and Wood, Klein, and other friends.

The markings on Venus and Jupiter as represented on pages 150 and 180 have come out much darker than was intended, but these illustrations may have some value as showing the position and form of the features delineated. It is difficult to reproduce delicate planetary markings in precisely the same characters as they are displayed in a good telescope. The apparent orbits of the satellites of the planets, delineated in figs. 41, 44, &c., are liable to changes depending on their variable position relatively to the Earth, and the diagrams are merely intended to give a good idea of these satellite systems.

W. F. D.

Bishopston, Bristol,
1891.

Omission, p. 220.—A column giving the periods of the satellites of Uranus should be added to the table as follows :—

d	h	m
2	12	29
4	3	27
8	16	57
13	11	7

PLATES I. and II. are views of the Observatory and Instruments recently erected by Mr. Klein at Stanmore, Middlesex, lat. 51° 36′ 57″ N., long. 0° 18′ 22″ W. The height above sea-level is 262 feet. The telescope is a 20-inch reflector by Calver, of 92 inches focus ; the tube is, however, 152 inches long so as to cut off all extraneous rays. It is mounted equatoreally, and is provided with a finder of 6 inches aperture—one of Tulley's famous instruments a century ago. The large telescope is fixed on a pillar of masonry 37 feet high, and weighing 115 tons. Mr. Klein proposes to devote the resources of his establishment to astronomical photography, and it has been provided with all the best appliances for this purpose. The observatory is connected by telephone with Mr. Klein's private residence, and the time-pieces and recording instruments are all electrically connected with a centre of observation in his study.

CONTENTS.

CONTENTS.

CHAPTER IX.

CHAPTER X.

CHAPTER XI.

CHAPTER XII.

CHAPTER XIII.

CHAPTER XIV.

CHAPTER XV.

CHAPTER XVI.

CHAPTER XVII.

ILLUSTRATIONS.

TELESCOPIC WORK

FOR

STARLIGHT EVENINGS.

CHAPTER I.

THE TELESCOPE, ITS INVENTION AND THE DEVE-LOPMENT OF ITS POWERS.

THE instrument which has so vastly extended our knowledge of the Universe, which has enabled us to acquire obser-vations of remarkable precision, and supplied the materials for many sublime speculations in Astronomy, was invented early in the seventeenth century. Apart from its special application as a means of exploring the heavens with a capacity that is truly marvellous, it is a construction which has also been utilized in certain other departments with signal success. It provided mankind with a medium through which to penetrate far beyond the reach of natural vision, and to grasp objects and phenomena which had either eluded detection altogether or had only been seen in dim and uncertain characters. It has also proved a very efficient instrument for various minor purposes of instruction and recreation. The invention of the telescope formed a new era in astronomy; and though, with a few exceptions, men were slow at first in availing themselves of its far-seeing resources, scepticism was soon swept aside and its value became widely acknowledged.

B

But though the telescope was destined to effect work of the utmost import, and to reach a very high degree of excellence in after times, the result was achieved gradually. Step by step its powers were enlarged and its qualities perfected, and thus the stream of astronomical discovery has been enabled to flow on, stimulated by every increase in its capacity.

There is some question as to whom may be justly credited with the discovery of its principles of construction. Huygens, in his ' Dioptrics,' remarks :—" I should have no hesitation in placing above all the rest of mankind the individual who, solely by his own reflections, without the aid of any fortuitous circumstances, should have achieved the invention of the telescope." There is reason to conclude, however, that its discovery resulted from accident rather than from theory. It is commonly supposed that Galileo Galilei is entitled to precedence ; but there is strong evidence to show that he had been anticipated. In any case it must be admitted that Galilei* had priority in successfully utilizing its resources as a means of observational discovery ; for he it was who, first of all men, saw Jupiter's satellites, the crescent form of Venus, the mountains and craters on the Moon, and announced them to an incredible world.

It has been supposed, and not without some basis of probability, that a similar instrument to the telescope had been employed by the ancients ; for certain statements contained in old historical records would suggest that the Greek philosophers had some means of extending their knowledge further than that permitted by the naked eye. Democritus remarked that the Galaxy or " Milky Way " was nothing but an assemblage of minute stars ; and it has been asked, How could he have derived this information but by instrumental aid ? It is very probable he gained the knowledge by inferences having their source in close observation ; for anyone who attentively studies the face of the sky must be naturally led to conclude that the appearance of the " Milky Way " is induced by immense and irregular clus-

* Galileo Galilei is very generally called by his christian name, but I depart from this practice here.

terings of small stars. In certain regions of the heavens there are clear indications of this : the eye is enabled to glimpse some of the individual star-points, and to observe how they blend and associate with the denser aggregations which give rise to the milky whiteness of the Galaxy.

Refracting lenses, or "burning-glasses," were known at a very early period. A lens, roughly figured into a convex shape and obviously intended for magnifying objects, has been recovered from the ruins of Herculaneum, buried in the ejections from Vesuvius in the year 79 A.D. Pliny and others refer to lenses that burnt by refraction, and describe globules of glass or crystal which, when exposed in the sun, transmit sufficient heat to ignite combustible material. The ancients undoubtedly used tubes in the conduct of their observations, but no lenses seem to have been employed with them, and their only utility consisted in the fact of their shutting out the extraneous rays of light. But spectacles were certainly known at an early period. Concave emeralds are said to have been employed by Nero in witnessing the combats of the gladiators, and they appear to have been the same in effect as the spectacles worn by short-sighted people in our own times. But the ancients supposed that the emerald possessed inherent qualities specially helpful to vision, rather than that its utility resulted simply from its concavity of figure. In the 13th century spectacles were more generally worn, and the theory of their construction understood.

It is remarkable that the telescope did not come into use until so long afterwards. Vague references were made to such an instrument, or rather suggestions as to the possibilities of its construction, which show that, although the principle had perhaps been conceived, the idea was not successfully put into practice. Roger Bacon, who flourished in the 13th century, wrote in his ' *Opus Majus* ' :—" Greater things may be performed by refracted light, for, from the foregoing principles, follows easily that the greatest objects may be seen very small, the remote very near, and *vice versâ*. For we can give transparent bodies such form and position with respect to the eye and the object that the rays are refracted and bent to where we like, so that we, under any angle, see the objects near or far, and in that manner we can, at a great distance, read the smallest letters, and we can count atoms and sand-grains, on account of the greatness of the angle under which they are seen."

Fracastor, in a work published at Venice in 1538, states:—" If we look through two eye-lenses, placed the one upon the other, everything will appear larger and nearer." He also says:—"There are made certain eye-lenses of such a thickness that if the moon or any other celestial body is viewed through them they appear to be so near that their distance does not exceed that of the steeples of public buildings."

In other writings will also be found intimations as to the important action of lenses; and it is hardly accountable that a matter so valuable in its bearings was allowed to remain without practical issues. The progressive tendency and the faculty of invention must indeed have been in an incipient stage, and contrasts strongly with the singular avidity with which ideas are seized upon and realized in our own day.

Many important discoveries have resulted from pure accident; and it has been stated that the first *bonâ fide* telescope had its origin in the following incident:—The children of a spectacle-maker, Zachariah Jansen, of Middleberg, in Zealand, were playing with some lenses, and it chanced that they arranged two of them in such manner that, to their astonishment, the weathercock of an adjoining church appeared much enlarged and more distinct. Having mentioned the curious fact to their father, he immediately turned it to account, and, by fixing two lenses on a board, produced the first telescope!

This view of the case is, however, a very doubtful one, and the invention may with far greater probability be attributed to Hans Lippersheim in 1608. Galilei has little claim to be considered in this relation; for he admitted that in 1609 the news reached him that a Dutchman had devised an appliance capable of showing distant objects with remarkable clearness. He thereupon set to work and experimented with so much aptitude on the principles involved that he very soon pro-'duced a telescope for himself. With this instrument he detected the four satellites of Jupiter in 1610, and other successes shortly followed. Being naturally gratified with the improvements he had effected in its construction, and with the wonderful discoveries he had made by its use, we can almost excuse the enthusiasm which prompted him to attribute the invention to his own ingenuity. But while according him the honour due to his sagacity in devoting this instrument to such excellent work, we must not overlook the fact that his claim to priority cannot be justified. Indeed, that Galilei had usurped the title of inventor is mentioned in letters which passed between the scientific men of that time. Fuccari, writing to Kepler, says :—" Galileo wants to be considered the inventor of the telescope, though he, as well as I and others, first saw the telescope which a certain Dutchman first brought with him to Venice, and although he has only improved it very little."

In a critical article by Dr. Doberck *, in which this letter is quoted and the whole question reviewed with considerable care, it is stated that Hans Lippersheim (also known as Jan Lapprey), who was born in Wesel, but afterwards settled at Middleberg, in the Netherlands, as a spectacle-maker, was really the first to make a telescope, and the following facts are quoted in confirmation:—" He solicited the States, as early as the 2nd October, 1608, for a patent for thirty years, or an annual pension for life, for the instrument he had invented, promising then only to construct such instruments for the Government. After inviting the inventor to improve the instrument and alter it so that they could look through it with both eyes at the same time, the States determined, on the 4th October, that from every province one deputy should be elected to try the apparatus and make terms with him concerning the price. This committee declared on the 6th October that it found the invention useful for the country, and had offered the inventor 900 florins for the instrument. He had at first asked 3000 florins for three instruments of rock-crystal. He was then ordered to deliver the instrument within a certain time, and the patent was promised him on condition that he kept the invention secret. Lapprey delivered the instrument in due time. He had

* ' Observatory,' vol. ii. p. 364.

arranged it for both eyes, and it was found satisfactory; but they forced him, against the agreement, to deliver two other telescopes for the same money, and refused the patent because it was evident that already several others had learned about the invention."

The material from which the glasses were figured appears to have been quartz; and efforts were made to keep the invention a profound secret, as it was thought it would prove valuable for "strategetical purposes." The cost of these primitive binoculars was about £75 each.

It is singular that, after being allowed to rest so long, the idea of telescopic construction should have been carried into effect by several persons almost simultaneously, and that doubts and disputes arose as to precedence. The probable explanation is that to one individual only priority was really due, but that, owing to the delays, the secret could not be altogether concealed from two or three others who recognized the importance of the discovery and at once entered into competition with the original inventor. Each of these fashioned his instrument in a slightly different manner, though the principle was similar in all; and having in a great measure to rely upon his individual faculties in completing the task, he considered himself in the light of an inventor and put forth claims accordingly. Not only were attempts made to assume the position of inventor, but there arose fraudulent claimants to some of the discoveries which the instrument effected in the hands of Galilei. Simon Marius, himself one of the very first to construct a telescope and apply it to the examination of the heavenly bodies, asserted that he had seen the satellites of Jupiter on December 29, 1609, a few days before Galilei, who first glimpsed them on January 7, 1610. Humboldt, in his 'Physical Description of the Heavens,' definitely ascribes the discovery of these moons to Marius; but other authorities uniformly reject the statement, and accord to Galilei the full credit.

It is stated that Galilei's first instrument magnified only three times, but he so far managed to amplify its resources that he was ultimately enabled to apply a power of 30. The lenses consisted of a double-convex object-glass, and a small double-concave eye-glass placed in front of the focal image formed by the object-glass. The ordinary opera-glass is constructed on a similar principle.

Fig. 1.

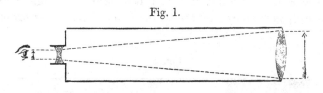

The Galilean Telescope.

The discoveries which Galilei effected with this crude and defective instrument caused a great sensation at the time. He made them known through the medium of a publication which he issued under the title of ' *Nuncius Siderus,*' or ' The Messenger of the Stars.' In that superstitious age great ignorance prevailed, bigotry was dominant, and erroneous views of the solar system were upheld and taught by authority. We can therefore readily conceive that Galilei's discoveries, and the direct inferences he put upon them, being held antagonistic to the ruling doctrines, would be received with incredulity and opposition. His views were regarded as heretical. In consequence of upholding the Copernican system he suffered persecution, and had to resort to artifice in the publication of his works. But the marvels revealed by his telescope, though discredited at first, could not fail to meet with final acceptance, for undeniable testimony to their reality was soon forthcoming. They were not, however, regarded until long afterwards as affirming the views enunciated by their clever author. Ultimately the new astronomy, based on the irrepressible evidence of the telescope, and clad in all the habiliments of truth, took the place of the old fallacious beliefs, to form an enduring monument to Copernicus and Galilei, who spent their lives in advancing its cause.

No special developments in the construction of the telescope appear to have taken place until nearly half a century subsequent to its invention. Kepler suggested an instrument formed of two convex lenses, and Scheiner and Huygens made telescopes on this principle in the middle of the 17th century. Huygens found great advantage in the employment of a compound eyepiece consisting of two convex lenses, which corrected the spherical aberration, and, besides being achromatic, gave a much larger field than the single lens. This eyepiece, known as the " Huygenian," still finds favour with the makers of telescopes.

Fig. 2.

Royal Observatory, Greenwich, in Flamsteed's time *.

Huygens may be said to have inaugurated the era of *long* telescopes. He erected instruments of 12 and 23 feet, having an aperture of 2⅓ inches and powers of 48, 50, and 92. He afterwards produced one 123 feet in focal length and 6 inches in aperture. Chief among his discoveries were the largest satellite of Saturn (Titan) and the true form of Saturn's ring. Hevelius of Dantzic built an instrument 150 feet long, which he fixed to a mast 90 feet in height, and regulated by ropes and pulleys. Cassini, at the Observatory at Paris, had telescopes by Campani of 86,

* Reproduced, by permission, from Cassell's 'New Popular Educator.'

100, and 136 French feet in length ; but the highest powers
he used on these instruments do not appear to have exceeded
150 times.   He made such good use of them as to discover
three of the satellites of Saturn and the black division in the
ring of that planet.   The largest object-glasses employed by
Hevelius and Cassini were of 6, 7, and 8 inches diameter.
This was during the latter half of the 17th century.   In
1712 Bradley made observations of Venus, and obtained
measures of the planet's diameter, with a telescope no less
than 212 feet in focal length.   The instruments alluded to
were manipulated with extreme difficulty, and observations
had to be conducted in a manner very trying to the
observer.   Tubes were sometimes dispensed with, the object-
glass being fixed to a pole and its position controlled by
various contrivances—the observer being so far off, however,
that he required the services of a good lantern in order to
distinguish it !

The immoderate lengths of refracting-telescopes were
necessary, as partially avoiding the effects of chromatic
aberration occasioned by the different refrangibility of the
seven coloured rays which collectively make white light.   In
other words, the coloured rays having various indices of refraction cannot be brought to a coincident focus by transmission
through a single lens.   Thus the red rays have a longer focus
than the violet rays, and the immediate effect of the different
refractions becomes apparent in the telescopic images, which
are fringed with colour and not sharply defined.   High
magnifying powers serve to intensify the obstacle alluded
to, and thus the old observers found it imperative to employ
eye-glasses not beyond a certain degree of convexity.   The
great focal lengths of their object-lenses enabled moderate
power to be obtained, though the eye-glass itself had a focus
of several inches and magnified very little.

Sir Isaac Newton made many experiments upon colours,
and endeavoured to obviate the difficulties of chromatic aberration, but erroneously concluded that it was not feasible.
He could devise no means to correct that dispersion of colour
which, in the telescopes of his day, so greatly detracted from
their effectiveness.   His failure seems to have had a pre-

judicial effect in delaying the solution of the difficulty, which
was not accomplished until many years afterwards.

Fig. 3.

Sir Isaac Newton*.

The idea of reflecting-telescopes received mention as early
as 1639 ; but it was not until 1663 that Gregory described
the instrument, formed of concave mirrors, which still bears
his name. He was not, however, proficient in mechanics,

Fig. 4.

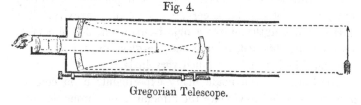

Gregorian Telescope.

and after some futile attempts to carry his theory into effect
the exertion was relinquished. In 1673 Cassegrain revived
the subject, and proposed a modification of the form pre-

* Reproduced, by permission, from Cassell's 'New Popular Educator.'

viously indicated by Gregory. Instead of the small concave mirror, he substituted a convex mirror placed nearer the speculum; and this arrangement, though it made the telescope shorter, had the disadvantage of displaying objects

Fig. 5.

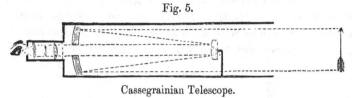

Cassegrainian Telescope.

in an inverted position. But the utility of these instruments was not demonstrated in a practical form until 1674, when Hooke, the clever mechanician, gave his attention to the subject and constructed the first one that was made of the kind.

In the meantime (1672) Sir Isaac Newton had completed with his own hands a reflecting-telescope of another pattern. In this the rays from the large concave speculum were received by a small plane mirror fixed centrally at the other

Fig. 6.

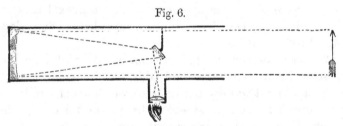

Newtonian Telescope.

end of the tube, and inclined at an angle of 45°; so that the image was directed at right angles through an opening in the side, and there magnified by the eye-lens. But for a long period little progress was effected in regard to reflecting-telescopes, owing to the difficulty of procuring metal well adapted for the making of specula.

In 1729 Mr. Chester Moor Hall applied himself to the study of refracting-telescopes and discovered that, by a combination of different glasses, the colouring of the images might be eliminated. It is stated that Mr. Hall made several achromatic glasses in 1733. A quarter of a century after this

John Dollond independently arrived at the same result, and took out a patent for achromatic telescopes. He found, by experiments with prisms, that crown and flint glass operated unequally in regard to the divergency of colours induced by refraction ; and, applying the principle further, he obtained a virtually colourless telescope by assorting a convex crown lens with a concave flint lens as the object-glass. Dollond also made many instruments having triple

Fig. 7.

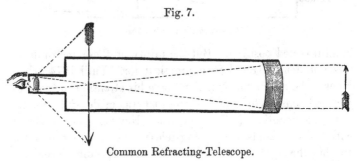

Common Refracting-Telescope.

object-lenses, and in these it was supposed that previous defects were altogether obliterated. Two convex lenses of crown glass were combined with a concave lens of flint glass placed between them.

Whether we regard Hall or Dollond as entitled to the most praise in connection with this important advance, it is certain that it was one the value of which could hardly be overestimated. It may be said to have formed a new era in practical astronomy. Instruments only 4 or 5 feet long could now be made equally if not more effective than those of 123 and 150 feet previously used by Huygens and Hevelius. All the troubles incidental to these long un-manageable machines now disappeared, and astronomers were at once provided with a handy little telescope capable of the finest performances.

Reflecting-telescopes also underwent marked improvements in the eighteenth century. Short, the optician, who died in 1768, was deservedly celebrated for the excellent instruments he made of the Gregorian form. Towards the latter part of the century William Herschel, by indomitable perseverance, figured a considerable number of specula. Some of these were mounted as Newtonians; others were employed in the form known as the "Front view," in which a second mirror is dispensed with altogether, and the rays from the large concave speculum are thrown to the side of the tube and

Fig. 8.

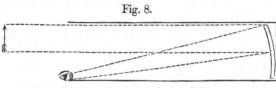

The Le Mairean or Herschelian Telescope.

direct to the eyepiece. This construction is often mentioned as the "Herschelian," but the idea had long before been detailed by Le Maire. In 1728 he presented a paper to the Académie des Sciences, giving his plans for a new reflecting-telescope. He proposed to suppress the small flat speculum in Newtonians, and " by giving the large concave speculum a little inclination, he threw the image, formed in its focus, to one side of the tube, where, an eyeglass magnifying it, the observer viewed it, his back at the time being turned towards the object in the heavens; thus the light lost in the Newtonian telescope by the second reflexion was saved."

After making several instruments of from 18 to 24 inches aperture, Herschel began one of larger calibre, and it was finished on August 28, 1789. The occasion was rendered historical by the discovery of one of the faintest interior satellites of Saturn, Enceladus. The large telescope had a speculum 48 inches in diameter; the tube was made of rolled or sheet iron, and it was 39 ft. 4 in. long and 4 ft. 10 in. in diameter. It was by far the largest instrument the world had seen up to that time; but it cannot be said to have realized the expectations formed of its powers, for its defining properties were evidently not on a par with its space-penetrating power. Many of Herschel's best observations were made with much smaller instruments. The large telescope, which was mounted in Herschel's garden at Slough, soon fell into comparative disuse, and, regarding it as incapable of further usefulness, Sir John Herschel sealed it up on January 1, 1840.

During the next half-century we hear of no attempts being made to surpass the large instrument which formed one of the working-tools of Herschel. Then, however, Lord Rosse entered the field, and in the 'Philosophical Transactions' for 1840 described a reflector of 3-feet diameter which he had set up at his residence at Parsonstown, Ireland. In 1845 the same nobleman, distinguished alike for his scientific attainments as for his generosity and urbanity of disposition, erected another telescope, the large speculum of which was 6 feet in diameter, $5\frac{1}{2}$ inches in thickness, and its weight 3 tons. Lord Rosse subsequently cast a duplicate speculum of 6 feet and weighing 4 tons. In point of dimensions this instrument far exceeded that of Herschel, and it is still in use, retaining its character as the largest, though certainly not the best, telescope in existence. Its tube is made of 1-inch deal, well bound together with iron hoops; it is 56 feet long and 7 feet in diameter.

Mr. Lassell soon afterwards made large specula. He erected one of 2-feet aperture and 20-feet focus at his residence at Starfield, near Liverpool, and in 1861 mounted one of 4-feet diameter and 37-feet focus. This instrument was for some time usefully employed by him at Malta. After Mr. Lassell's return to England his great telescope remained in a dismantled state for several years, and ultimately the speculum was broken up and " consigned to the crucible of the bell-founder."

It is not a little remarkable that Herschel, Rosse, and Lassell personally superintended and assisted in the construction of the monster instruments with which their names are so honourably associated.

In or about the year 1867 a telescope of the Cassegrainian form, and having a metallic speculum 4 feet in diameter and 28-feet focus, was completed by Grubb of Dublin for the observatory at Melbourne. This instrument, which cost something like £14,000, was found defective at first, though the fault does not appear to have rested with the optician.

Up to this period specula were formed of a metal in which copper and tin were largely represented. But the days of metal specula were numbered. Leon Foucault, in the year

1859, published a valuable memoir in which he described
the various ingenious methods he employed in figuring surfaces of glass to the required curve. He furnished data for
determining accuracy of figure. Formerly opticians had
considerable trouble in deciding the quality of their newly-
ground specula or object-glasses. They found it expedient to
mount them temporarily, and then, by actual trial on difficult
objects, to judge of their efficiency. This involved labour and
occasioned delay, especially in the case of large instruments.
Foucault showed that crucial tests might be applied in the
workshop, and that glasses could be turned out of hand
without any misgivings as to their perfection of figure.

Foucault's early experiments in parabolizing glass led him
to important results. By depositing a thin coating of silver
on his specula he obtained a reflective power far surpassing
that of metal. Thereafter metal was not thought of as a
suitable material for reflecting-telescopes. Silver-on-glass
mirrors immediately came into great request. The latter
undoubtedly possess a great superiority over metal, especially as regards light-grasping power, the relative capacity
according to Sir J. Herschel being as ·824 to ·436. Glass
mirrors have also another advantage in being less heavy than
those of metal. It is true the silver film is not very durable,
but it can be renewed at any time with little trouble or
expense.

With of Hereford, and after him Calver of Chelmsford,
became noted for the excellency of their glass mirrors. They
were found nearly comparable to refractors of the same
aperture.

A tendency of the times was evidently in the direction
of large instruments. One of 47·2-inches aperture (for
which a sum of 190,000 francs was paid) was completed
by Martin in 1875 for the Paris Observatory, but its employment since that year has not furnished a very successful
record. The largest instrument of the kind yet made has
a speculum 5 feet in diameter and 27½-feet focal length. It
was placed in position in September 1888, and was made
by the owner, Mr. Common, of Ealing, whose previous
instrument was a 37-inch glass reflector by Calver. The

5-foot telescope is undoubtedly of much greater capacity than the colossal reflector of Lord Rosse, though it is not so large.

Mr. Calver has recently figured a 50-inch mirror for Sir H. Bessemer, but the mounting is not completed; and he is expecting to make other large reflectors, viz. one over 5 feet in diameter and another over 3 feet. The late Mr. Nasmyth also erected some fine instruments, and adopted a combination of the Cassegrainian and Newtonian forms to ensure greater convenience for the observer. Instead of permitting the rays from the small convex mirror to return through the large mirror, he diverted them through the side of the tube by means of a flat mirror, as in Newtonians. But this construction is not to be commended, because much light is lost and defects increased by the additional mirror.

Smaller telescopes of the kind we have been referring to have become extremely popular : and deservedly so. They are likely to maintain their character in future years; for the Newtonian form of instrument, besides being thoroughly effective in critical work, is moderate in price and gives images absolutely achromatic. Moreover, it is used with a facility and ease which an experienced observer knows how to appreciate. Whatever may be the altitude of the objects under scrutiny, he is enabled to retain a perfectly convenient and natural posture, and may pursue his work during long intervals without any of the fatigue or discomfort incidental to the use of certain other forms of instrument.

Returning now to refractors : many years elapsed after Dollond patented his achromatic object-glass before it was found feasible to construct these instruments of a size sufficient to grasp faint and delicate objects. Opticians were thwarted in their efforts to obtain glass of the requisite purity for lenses, unless in small disks very few inches in diameter. It is related that Dollond met with a pot of uncommonly pure flint glass in 1760, but even with this advantage of material he admitted that, after numerous attempts, he could not provide really excellent object-glasses of more than $3\frac{3}{4}$-inches diameter. It may therefore be readily imagined that a refractor of $4\frac{1}{2}$ or 5-inches aperture was an instrument

Fig. 9.

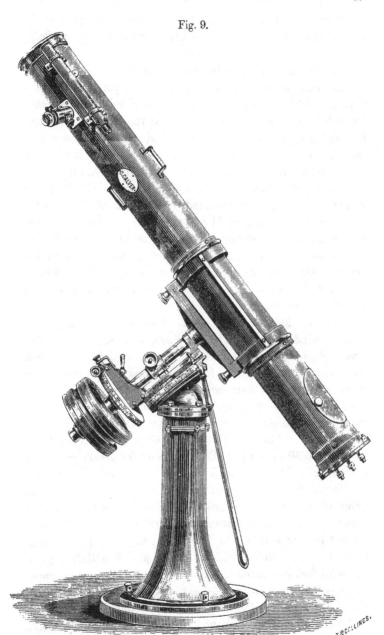

10-inch Reflecting-Telescope on a German Equatoreal, by Calver.

of great rarity and expense. Towards the latter part of the 18th century Tulley's price was £275 for a 5-inch equatoreally mounted.

In later years marked improvements were effected in the manufacture of glass. A sign of this is apparent in the fact that, in 1829, Sir James South was enabled to purchase a 12-inch lens. Four years before this the Dorpat telescope, having an objective of 9½ inches, had created quite a sensation. As time went on, still larger glasses were made. In 1862 Alvan Clark & Sons, of New York, U.S.A., finished an instrument of 18½-inches aperture, at a cost of £3700 ; and in 1869 Cooke & Sons mounted a 24·6-inch object-glass for the late Mr. Newall, of Gateshead. The latter instrument was much larger than any other refractor hitherto made, but it was not long to maintain supremacy. One of 25·8 inches and 29-feet focus was finished in 1872 by Alvan Clark & Sons for the Naval Observatory, Washington, at a cost of £9000. Another, of similar size, was supplied by the same firm to Mr. McCormick, U.S.A. Several important discoveries, including the satellites of Mars, were effected with the great Washington telescope. A few years later a 27-inch was completed by Grubb for the Vienna Observatory, and quite recently the four largest refractors ever made have been placed in position and are actively employed in various departments of work. These include a 29-inch by Martin for the Paris Observatory, a 30-inch by Henry Bros. for Nice, a 30-inch by A. Clark & Sons for Pulkowa, and a 36-inch, also by A. Clark & Sons, for the Lick Observatory on Mount Hamilton in California. The latter has no rival in point of size, though rumours are current that still larger lenses are in contemplation. The tube of the 36-inch is 56 feet long and 3½ feet in diameter at the ends, but the diameter is greater in the middle. It is placed within a great dome 75 feet in diameter. The expense of the entire apparatus is given as follows :—Cost of the dome, $56,850 ; of the visual objective, $53,000 ; of the photographic objective, $13,000 ; of the mounting, $42,000. Total, $164,850. This noble instrument—due to the munificence of one individual, the late Mr. James Lick, of Chicago, who bequeathed

$700,000 for the purpose—may be regarded as the king of refracting-telescopes. Placed on the summit of Mount Hamilton, where the atmosphere is exceptionally favourable for celestial observations, and utilized as its resources are by some of the best observers in America, we may confidently expect it to largely augment our knowledge of the heavenly bodies.

The great development in the powers of both refracting and reflecting-telescopes, as a means of astronomical discovery, exemplifies in a remarkable degree the ever-increasing resources and refinements of mechanical art. In 1610 Galilei, from his window at Padua, first viewed the moon and planets with his crude instrument having a power of 3, and he achieved much during the remaining years he lived, by increasing it tenfold, so that at last he could magnify an object 30 times. Huygens laboured well in the same field ; and others who succeeded him formed links in the chain of progress which has almost uninterruptedly run through all the years separating Galilei's time from our own. The primitive efforts of the Florentine philosopher appear to have had their sequel in the magnificent telescope which has lately been erected under the pure sky of Mount Hamilton. The capacity of this instrument relatively to that of earlier ones may be judged from the fact that a power of about 3300 times has lately been employed with success in the measurement of a close and difficult double star. Could Galilei but stand for a few moments at the eyepiece of this great refractor, and contemplate the same objects which he saw, nearly three centuries ago, through his imperfect little glasses at Padua, he would be appalled at the splendid achievements of modern science.

CHAPTER II.

RELATIVE MERITS OF LARGE AND SMALL TELESCOPES.

THE number of large telescopes having so greatly increased in recent years, and there being every prospect that the demand for such instruments will continue, it may be well to consider their advantages as compared with those of much inferior size. Object-glasses and specula will probably soon be made of a diameter not hitherto attained; for it is palpably one of the ambitions of the age to surpass all previous efforts in the way of telescopic construction. There are some who doubt that such enormous instruments are really necessary, and question whether the results obtained with them are sufficient return for the great expense involved in their erection. Large instruments require large observatories; and the latter must be at some distance from a town, and in a locality where the atmosphere is favourable. Nothing can be done with great aperture in the presence of smoke and other vapours, which, as they cross the field, become ruinous to definition. Moreover, a big instrument is not to be manipulated with the same facility as a small one : and when anything goes wrong with it, its rectification may be a serious matter, owing to the size. Such telescopes need constant attention if they would be kept in thorough working order. On the other hand, small instruments involve little outlay, they are very portable, and require little space. They may be employed in or out of doors, according to the inclination and convenience of the observer. They are controlled with the greatest ease, and seldom get out of adjustment. They are less susceptible to atmospheric influences than larger instruments, and hence may be used more frequently with success and at places by no means favourably situated in this respect. Finally, their defining powers are of such excellent character as to compensate in a measure for feeble illumination.

In discussing this question it will be advisable to glance at the performances of certain instruments of considerable size.

The introduction of really large glasses dates from a century ago, when Sir W. Herschel mounted his reflector, 4 feet in aperture, at Slough. He discovered two of the inner satellites of Saturn very soon after it was completed ; but apart from this the instrument seems to have achieved little. Herschel remarked that on August 28, 1789, when he brought the great instrument to the parallel of Saturn, he saw the spots upon the planet better than he had ever seen them before. The night was probably an exceptionally good one, for we do not find this praise reiterated. Indeed, Herschel appears to have practically discarded his large instrument for others of less size. He found that with his small specula of 7-ft. focus and 6·3-in. aperture he had "light sufficient to see the belts of Saturn completely well, and that here the maximum of distinctness might be much easier obtained than where large apertures are concerned." Even in his sweeps for nebulæ he employed a speculum of 20-ft. focus and 18½-in. aperture in preference to his 4-ft. instrument, though on objects of this nature light-grasping power is essentially necessary. The labour and loss of time involved in controlling the large telescope probably led to its being laid aside for more ready means, though Herschel was not the man to spare trouble when an object was to be gained. His life was spent in gleaning new facts from the sky ; and had the 4-foot served his purpose better than smaller instruments, no trifling obstacle would have deterred him from its constant employment. But his aim was to accomplish as much as possible in every available hour when the stars were shining, and experience doubtless taught him to rely chiefly upon his smaller appliances as being the most serviceable. The Le Mairean form, or " Front view," which Herschel adopted for the large instrument may quite possibly have been in some degree responsible for its bad definition.

Lord Rosse's 6-ft. reflector has now been used for nearly half a century, and its results ought to furnish us with good evidence as to the value of such instruments. It has done important work on the nebulæ, especially in the reobservation of the objects in Sir J. Herschel's Catalogues of 1833 and 1864. To this instrument is due the discovery of spiral

Fig. 10.

Lord Rosse's 6-foot Reflecting-Telescope.

nebulæ ; and perhaps this achievement is its best. But when we reflect on the length of its service, we are led to wonder that so little has been accomplished. For thirty years the satellites of Mars eluded its grasp, and then fell a prize to one of the large American telescopes. The bright planets * have been sometimes submitted to its powers, and careful drawings

* Such objects show considerable glare in a very large instrument. The advent of Jupiter into the field of the 6-foot has been compared to the brightness of a coach-lamp. The outer satellite of Mars was seen twice with this instrument in 1877, "but the glare of the planet was found too strong to allow of good measures being taken."

executed by good observers; but they show no extent of
detail beyond what may be discerned in a small telescope.
This does not necessarily impugn the figure of the large
speculum, the performance of which is entirely dependent
upon the condition of the air. The late Dr. Robinson, of
Armagh, who had the direction of the instrument for some
time, wrote in 1871:—"A stream of heated air passing
before the telescope, the agitation and hygrometric state of
the atmosphere, and any differences of temperature between
the speculum and the air in the tube are all capable of
injuring or even destroying definition, though the speculum
were absolutely perfect. The effect of these disturbances is,
in reflectors, as the cube of their apertures; and hence there
are few hours in the year when the 6-foot can display its full
powers." Another of the regular observers, Mr. G. J.
Stoney, wrote in 1878:—"The usual appearance [of the
double star γ^2 Andromedæ] with the best mirrors was a
single bright mass of blue light some seconds in diameter
and boiling violently." On the best nights, however, "the
disturbance of the air would seem now and then suddenly to
cease for perhaps half a second, and the star would then
instantly become two very minute round specks of white
light, with an interval between which, from recollection, I
would estimate as equal to the diameter of either of them, or
perhaps slightly less. The instrument would have furnished
this appearance uninterruptedly if the state of the air had
permitted." The present observer in charge, Dr. Boeddicker,
wrote the author in 1889:—"There can be no doubt that on
favourable nights the definition of the 6-foot is equal to that
of any instrument, as is fully shown by Dr. Copeland's
drawings of Jupiter published in the 'Monthly Notices' for
March 1874. It appears to me, however, that the advantage
in going from the 3-foot to the 6-foot is not so great in the
case of planets as in the case of nebulæ; yet, as to the Moon,
the detail revealed by the 6-foot on a first-class night is
simply astounding. The large telescope is a Newtonian
mounted on a universal joint. For the outlying portions of
the great drawing of the Orion nebula it was used as a
Herschelian. As to powers profitably to be used, I find

no advantage in going beyond 600 ; yet formerly on short
occasions (not longer than perhaps 1 hour a night) very much
higher powers (over 1000) have been successfully employed
by my predecessors."

Mr. Lassell's 4-foot reflector was taken to Malta, and while
there its owner, assisted by Mr. Marth, discovered a large
number of nebulæ with it, but it appears to have done nothing
else.   His 2-foot reflector, which he had employed in
previous years, seems to have been his most effective instrument ; for with this he discovered Ariel and Umbriel, the
two inner satellites of Uranus, Hyperion, the faintest satellite
of Saturn, and the only known satellite of Neptune.   He also
was one of the first to distinguish the crape ring of Saturn.
Mr. Lassell had many years of experience in the use of large
reflectors ; and in 1871 he wrote :—" There are formidable
and, I fear, insurmountable difficulties attending the construction of telescopes of large size. . . . These are, primarily,
the errors and disturbances of the atmosphere and the flexure
of the object-glasses or specula.   The visible errors of the
atmosphere are, I believe, generally in proportion to the
aperture of the telescope. . . . Up to the size [referring to
an 8-in. O.-G.] in question, seasons of tranquil sky may be
found when its errors are scarcely appreciable ; but when we
go much beyond this limit (say to 2 feet and upwards), both
these difficulties become truly formidable.   It is true that the
defect of flexure may be in some degree eliminated, but that
of atmospheric disturbance is quite unassailable.   These circumstances will always make large telescopes *proportionately*
less powerful than smaller ones ; but notwithstanding these
disadvantages they will, on some heavenly objects, reveal
more than any small ones can."   Mr. Lassell's last sentence
refers to " delineations of the forms of the fainter nebulæ,"
to " seeing the inner satellites of Uranus, the satellite of
Neptune, and the seventh satellite of Saturn."   He mentions
that, when at Malta, he " saw, in the 2-foot equatoreal, with
a power of 1027, the two components of γ^2 Andromedæ
distinctly separated to the distance of a neat diameter of the
smaller one.   Now, no telescope of anything like 8-inches
diameter could exhibit the star in this style."

The large Cooke refractor of 24·8-inches aperture, which has been mounted for about twenty years at Gateshead, has a singularly barren record. Its atmospheric surroundings appear to have rendered it impotent. The owner of this fine and costly instrument wrote the author in 1885:—"Atmosphere has an immense deal to do with definition. I have only had one fine night since 1870! I then saw what I have never seen since."

The Melbourne reflector of 4-feet aperture performed very indifferently for some years, and little work was accomplished with it. Latterly its performance has been more satisfactory; excellent photographs of the Moon have been taken, and it has been much employed in observations of nebulæ. The speculum having recently become tarnished, it has been dismounted for the purpose of being repolished.

The silver-on-glass reflector of 47·2-in. diameter, at the Paris Observatory, was used for some years by M. Wolf, who has also had the control of smaller telescopes. He was in a favourable position to judge of their relative effectiveness. In a lecture delivered at the Sardonne on March 6, 1886, he said:—"During the years I have observed with the great Parisian telescope I have found but one solitary night when the mirror was perfect." Further on, he adds:—"I have observed a great deal with the two instruments [both reflectors] of 15·7 inches and 47·2 inches. I have rarely found any advantage in using the larger one when the object was sufficiently luminous." M. Wolf also avers that a refractor of 15 inches or reflector of 15·7 inches will show everything in the heavens that can be discovered by instruments of very large aperture. He always found a telescope of 15·7-inch aperture surpass one of 7·9 inches, but expresses himself confidently that beyond about 15 inches increased aperture is no gain.

The Washington refractor of 25·8 inches effected a splendid success in Prof. Hall's hands in 1877, when it revealed the two satellites of Mars. But immediately afterwards these minute bodies were shown in much smaller instruments; whence it became obvious that their original discovery was not entirely due to the grasp of the 25·8-inch telescope, but

in a measure to the astuteness displayed by Prof. Hall in the
search. A good observer had been associated with a good
telescope ; and an inviting research having been undertaken,
it produced the natural result—an important success. The
same instrument, in the same hands, enabled the rotation-
period of Saturn to be accurately determined by means of a
white spot visible in December 1876 on the disk of the
planet, and which was subsequently seen by other observers
with smaller glasses. Good work in other directions has also
been accomplished at Washington, especially in observations
of double stars and faint satellites. But notwithstanding
these excellent performances, Prof. Hall expressed himself in
rather disparaging terms of his appliances, saying "the large
telescope does not show enough detail." He gave a more
favourable report in 1888 ; for we find it stated that " the
objective retains its figure and polish well. By comparison
with several other objectives which Prof. Hall has had an
opportunity of seeing during recent years, he finds that the
glass is an excellent one."

Prof. Young, who has charge of the 23-inch refractor
at Princeton, has also commented on the subject of the
definition of large telescopes. He says:—" The greater
susceptibility of large instruments to atmospheric disturbances is most sadly true ; and yet, on the whole, I find
also true what Mr. Clark told me would be the case on first
mounting our 23-inch instrument, that *I can almost always
see with the 23-inch everything I see with the 9½-inch under the
same atmospheric conditions, and see it better,*—if the seeing
is bad only a little better, if good immensely better." Prof.
Young also mentioned that a power of 1200 on the 23-inch
" worked perfectly on Jupiter on two different evenings in
the spring of 1885 in bringing out fine details relating to the
red spot and showing the true forms of certain white dots on
the S. polar belt."

The 26-inch refractor at the Leander McCormick Observatory, U.S.A., is successfully engaged in observations of
nebulæ, and many new objects of this character have been
found. It does not appear that the telescope is much used
for other purposes ; so that we can attach no significance to

the fact that important discoveries have not been made with it in other departments.

The great Vienna refractor of 27 inches aperture "does not seem to accomplish quite what was expected of it," according to Mr. Sawerthal, who ·recently visited the Observatory at Währing, Vienna. The Director, Dr. Weiss, states in his last report that " the 27-inch Grubb refractor has only been occasionally used, when the objects were too faint for the handier instruments."

The still larger telescopes erected at the Observatories at Pulkowa and Nice have so recently come into employment that it would be premature to judge of their performance. In the Annual Report from Pulkowa (1887) it is stated that Dr. H. Struve was using the 30-inch refractor "in measuring those of Burnham's double stars which are only seldom measurable with the 'old 15-inch,' together with other stars of which measures are scarce. He made 460 measures in eight or nine months, as well as 166 micrometric observations of the fainter satellites of Saturn and 15 of that of Neptune." At Nice the 30-inch refractor was employed by M. Perrotin in physical observations of Mars in May and June 1888. The canal-shaped markings of Schiaparelli were confirmed, and some of them were traced "from the ocean of the southern hemisphere right across both continents and seas up to the north polar ice-cap." The 30-inch also showed some remarkable changes in the markings ; but these were not confirmed at other observatories. The telescope evidently revealed a considerable amount of detail on this planet ; whence we may infer that its defining power is highly satisfactory.

The great Lick refractor, which appears to have been " first directed to the heavens from its permanent home on Mount Hamilton on the evening of January 3, 1888," has been found ample work by the zealous astronomers who have it in charge. Prof. Holden, in speaking of it, says :—" It needs peculiar conditions, but when all the conditions are favourable its performance is superb." Mr. Keeler, one of the observers, writes that, on January 7, 1888, when Saturn was examined, " he not only shone with the brilliancy due to the great size of the objective, but the minutest details of his surface were visible with wonderful distinctness. The outlines of the rings were very sharply defined with a power of 1000." Mr. Keeler adds :—" According to my experience, there is a direct gain in power with increase of aperture. The 12-inch equatoreal brings to view objects entirely beyond the reach of the $6\frac{1}{2}$-inch telescope, and details almost beyond perception with the 12-inch are visible at a glance with the 36-inch equatoreal. The great telescope is equal in defining power to the smaller ones." This is no small praise, and it must have been extremely gratifying, not only to those who were immediately associated with the construction of the telescope, but to astronomers everywhere who were hoping to hear a satisfactory report. In its practical results this instrument has not yet, it is true, given us a discovery of any magnitude. It has disclosed several very small stars in the trapezium of the Orion nebula, some difficult double stars have been found and measured, and some interesting work has been done on the planets and nebulæ. Physical details have been observed in the ring nebula, between β and γ Lyræ, which no other telescope has ever reached before.

Mr. Common's 5-foot reflector has been employed on several objects. In the spring of 1889 Uranus was frequently observed with it, and several minute points of light, suspected to be new satellites, were picked up. Evidence was obtained of a new satellite between Titania and Umbriel; but bad weather and haze, combined with the low altitude of Uranus, interfered with the complete success of the observations. " With only moderate powers, Uranus does not show a perfectly sharp disk. No markings are visible on it, and nothing like a ring has been seen round it." Mr. Common, in a letter to the writer, dated November 9, 1889, says :— " The 5-foot has only been tried in an unfinished state as yet, the mirror not being quite finished when put into the tube last year. This was in order to gain experience and save the season. It performed much better than I had hoped, and is greatly superior to the 3-foot. I took some very fine photographs with it last year. It has been refigured, or rather completed, this summer, and has just been resilvered." From this it is evident that Mr. Common's large instrument has not yet been fully tested ; but it clearly gives promise of successful results, and encourages the hope that it will exert an influence on the progress of astronomy. Owing to the highly reflective quality of silvered glass, the 5-foot speculum has a far greater command of light (space-penetrating power) than the great objective mounted at the Lick Observatory. Mr. Common's mirror may therefore be expected to grasp nebulæ, stars, satellites, and comets which are of the last degree of faintness and quite invisible in the Lick refractor. But we must not forget that the latter instrument is certainly placed in a better atmosphere, and that its action is not therefore arrested in nearly the same degree by haze and undulations of the air. With equal conditions, the great reflector at Ealing would probably far surpass the large refractor we have referred to, the latter having less than one third of the light-grasping power of the former.

This rapid sketch of the performances of some of our finest telescopes must suffice for the present in assisting us to estimate their value as instruments of discovery. And it must be admitted that, on the whole, these appliances have been disappointing. The record of their successes is by no means an extended one, and in some individual cases absolute failure is unmistakable. We must judge of large glasses by their revelations; their capacity must be estimated by results. We often meet with glowing descriptions of colossal telescopes : their advantages are specified and their performances extolled to such a degree that expectation is raised to the highest pitch. But it is not always that such praise is justified by facts. The fruit of their employment is rarely prolific to the extent anticipated, because the observers have been defeated in their efforts by impediments which inseparably attend the use of such huge constructions.

Our atmosphere is always in a state of unrest. Its condition is subject to many variations. Heat, radiated or evolved from terrestrial objects, rises in waves and floats along with the wind. These vapours exercise a property of refraction, with the result that, as they pass in front of celestial objects, the latter at once become subject to a rapid

series of contortions in detail. Their outlines appear tremulous, and all the features are involved in a rippling effect
that seriously compromises the definition. Delicate markings
are quite effaced on a disk which is thus in a state of ebullition ; and on such occasions observers are rarely able to
attain their ends. Telescopic work is, in fact, best deferred
until a time when the air has become more tranquil. In
large instruments these disturbances are very troublesome, as
they increase proportionately with aperture. They are so
pronounced and so persistent as to practically annul the
advantage of considerable light-grasping power ; for unless
the images are fairly well defined, mere brightness counts
for nothing. Reflectors are peculiarly susceptible to this
obstacle ; moreover, the open tube, the fact that rays from
an object pass twice through its length, and that a certain
amount of heat radiated from the observer must travel across
the mouth of the tube all serve to impair the definition. A
speculum, to act well, must be of coincident temperature in
every part. This is not always the case, owing to the
variableness of the weather or to unequal exposure of the
speculum. Large refractors, though decidedly less liable to
atmospheric influences, are yet so much at the mercy of them
that one of the first and most important things discussed in
regard to a new instrument is that of a desirable site for it.

The great weight of large objectives and specula tends
to endanger the perfect consistency and durableness of their
figure, and imposes a severe strain upon their cellular mounting. The glasses must obviously assume a variety of bearings
during active employment. This introduces a possible cause
of defective performance ; for in some instances definition
has been found unequal, according to the position of the
glass. Specula are very likely to be affected in this manner,
as they are loosely deposited in their cells to allow of expansion, and the adjustment is easily deranged. The slightest
flaw in the mounting of objectives immediately makes itself
apparent in faulty images. Special precautions are of course
taken to prevent flexure and other errors of the kind alluded
to, and modern adaptations may be said to have nearly eliminated them; but there is always a little outstanding danger,

from the ease with which glasses may be distorted or their adjustment become unsettled.

Another difficulty formerly urged against telescopes of great size was the trouble of managing them; but this objection can scarcely be applied to the fine instruments of the present day, which are so contrived as to be nearly as tractable as small ones. A century ago, glass of the requisite purity for large objectives could not be obtained; but this difficulty appears also to have quite disappeared. And the process of figuring lenses of considerable diameter is now effected with the same confidence and success as that of greatly inferior sizes.

Let us now turn for a moment to the consideration of small instruments, premising that in this category are included all those up to about 12-inches aperture. Modern advances have quite altered our ideas as to what may be regarded as large and small telescopes. Sixty-five years ago the Dorpat refractor, with a $9\frac{1}{2}$-inch objective by Fraunhofer, was considered a prodigy of its class; now it occupies a very minor place relatively to the 30-inch and 36-inch objectives at Nice, Pulkowa, and Mount Hamilton.

Prof. Hall remarked, in 1885:—"There is too much scepticism on the part of those who are observing with large instruments in regard to what can be seen with small ones." This is undoubtedly true; but a mere prejudice or opinion of this sort cannot affect the question we are discussing, as it is one essentially relying upon facts.

Small instruments have done a vast amount of useful work in every field of astronomical observation. Even in the realm of nebulæ, which, more than any other, requires great penetrating power, D'Arrest showed what could be effected with small aperture. Burnham, with only a 6-inch refractor, has equally distinguished himself in another branch; for he has discovered more double stars than any previous observer. Dawes was one of the most successful amateurs of his day, though his instrumental means never exceeded an 8-inch glass. But we need not particularize further. It will be best to get a general result from the collective evidence of past years. We find that nearly all the comets, planetoids,

Fig. 11.

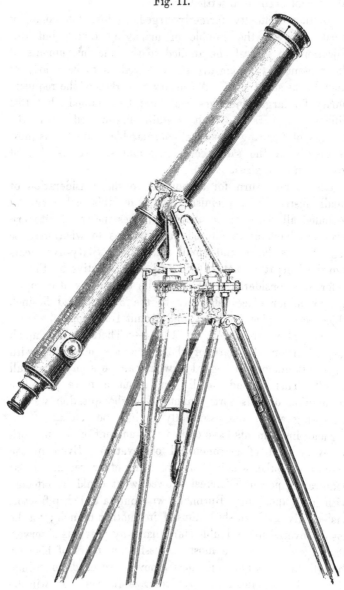

Refracting-Telescope, by Browning.

double stars, &c. owe their first detection to comparatively small instruments. Our knowledge of sun-spots, lunar and planetary features is also very largely derived from similar sources. There is no department but what is in lebted more or less to the services of small telescopes : the good work they have done is due to their excellent defining powers an to the facility with which they may be used.

We have already said that the record of discoveries mac with really large instruments is limited ; but it should also be remarked that until quite recently the number of such instruments has been very small. And not always, perhaps, have the best men had the control of them. Virtually the observer himself constitutes the most important part of his telescope : it is useless having a glass of great capacity at one end of a tube, and a man of small capacity at the other. Two different observers essentially alter the character of an instrument, according to their individual skill in utilizing its powers.

Large telescopes are invariably constructed for the special purpose of discovering unknown orbs and gleaning new facts from the firmament. But in attempting to carry out this design, obstacles of a grave nature confront the observer. The comparatively tranquil and sharply definite images seen in small instruments disappear, and in their places forms are presented much more brilliant and expansive, it is true, but involved in glare and subject to constant agitation, which serve to obliterate most of the details. The observer becomes conscious that what he has gained in light has been lost in definition. At times—perhaps on one occasion in fifty—this experience is different ; the atmosphere has apparently assumed a state of quiescence, and objects are seen in a great telescope with the same clearness of detail as in smaller ones. It is then the observer fully realizes that his instrument, though generally ineffective, is not itself in fault, and that it would do valuable work were the normal condition of the air suitable to the exercise of its capacity.

Those who have effected discoveries with large instruments have done so in spite of the impediment alluded to. Relying

D

mainly upon great illuminating power, bad or indifferent
definition has been tolerated ; and they have succeeded in
detecting minute satellites, faint nebulæ, clusters, and small
companions to double stars.   Telescopes of great aperture are
at home in this kind of work.   But when we come to consider discoveries on the surfaces of the Sun, Moon, and
planets, the case is entirely different ; the diligent use of
small appliances appears to have left little for the larger
constructions to do.   There are some thousands of drawings
of the objects named, made by observers employing telescopes
from 3 up to 72 inches in diameter ; and a careful inspection
shows that the smaller instruments have not been outdone in
this interesting field of observation.   In point of fact they
rather appear to have had the advantage, and the reason
of this is perhaps sufficiently palpable.   The details on a
bright planetary object are apt to become obliterated in the
glare of a large instrument.   Even with a small telescope
objects like Venus and Jupiter are best seen at about the
time of sunset, and before their excessive brilliancy on the
dark sky is enabled to act prejudicially in effacing the
delicate markings.   Probably this is one of the causes which,
in combination with the undulations of the atmosphere, have
restricted the discoveries of large instruments chiefly to faint
satellites, stars, and nebulæ.

Prof. Young ascribes many of the successes of small instruments to exceptional cuteness of vision on the part of certain
observers, and to the fact that such instruments are so very
numerous and so diligently used that it is fair to conclude
they must reap the main harvest of discoveries.   We must
remember that for every observer working with an aperture
of 18 inches and more, there are more than a hundred
employing objectives or specula of from 5 to 12 inches ;
hence we may expect some notable instances of keen sight
amongst the latter.   The success of men like Dawes and
others, who outstrip their contemporaries, and with small
glasses achieve phenomenal results, is to be ascribed partly to
good vision and partly to that natural aptitude and pertinacity uniformly characteristic of the best observers.   These
circumstances go far to explain the unproductiveness of large

telescopes : in the race for distinction they are often distanced by their more numerous and agile competitors.

The objections which applied to the large reflecting instruments of Herschel, Lassell, and Rosse scarcely operate with the same force in regard to the great refractors of the present day, and for these reasons :—Refractors are somewhat less sensitive to atmospheric disturbances than reflectors. The modern instruments are mounted in much improved style, and placed in localities selected for their reception. In fact, all that the optician's art can do to perfect such appliances has been done, and Nature herself has been consulted as to essentials ; for we find the most powerful refractor of all erected on the summit of Mount Hamilton, where the skies are clear and Urania ever smiles invitingly.

Some observers who have obtained experience both with large and small telescopes aver that, even on a bright planet, they can see more, and often see it much better, with the larger glasses. But we rarely, if ever, find them saying they can discern anything which is absolutely beyond the reach of small instruments. It would be much more satisfactory evidence of the super-excellence of the former if definite features could be detected which are quite beyond the reach of telescopes of inferior size ; but we seldom meet with experiences of this kind, and the inference is obvious.

There is undoubtedly a certain aperture which combines in itself sufficient light-grasping power with excellent definition. It takes a position midway between great illuminating power and bad definition on the one hand, and feeble illuminating power and sharp definition on the other. Such an aperture must form the best working instrument in an average situation upon ordinary nights and ordinary objects. M. Wolf fixes this aperture at about 15 inches, and he is probably near the truth.

The quaint Dr. Kitchener, who, early in the present century, made a number of trials with fifty-one telescopes, entertained a very poor opinion of big instruments. In his book on ' Telescopes,' he says :—" Immense telescopes are only about as useful as the enormous spectacles suspended over the doors of opticians." . . . "Astronomical amateurs should rather seek for *perfect* instruments than *large* ones. What good can a great deal of bad light do?"

We shall be in a better position a few years hence to estimate the value of great telescopes; for the principal instruments of this class have only been completed a short time. Judging from the statements of some of the observers, who are men of the utmost probity and ability, certain of the large instruments are capable of work far in advance of anything hitherto done. Definition, they say, is excellent, notwithstanding the great increase of aperture. The old stumbling-block appears, therefore, to have been removed, and astronomy is to be congratulated on the acquirement of such vastly improved implements of research. Even should the large telescopes continue to prove disappointing in certain branches, they may certainly be expected to maintain their advantage in others. They will always be valuable as a corrective to smaller and handier instruments. For special lines of work in which very small or very faint objects are concerned, considerable light-grasping power is absolutely required; and it is chiefly in these departments that large instruments may be further expected to augment our knowledge. In photographic and spectroscopic work they also have a special value, which late researches have brought prominently to the fore.

The telescopes of the future will probably surpass in dimensions those of our own day. The University of Los Angelos, in California, propose to erect a 42-inch refractor on the summit of Wilson's Peak of the Sierra Madre mountains, which is 6000 feet high and about 25 miles from Los Angelos. In reference to this contemplated extension of size, it may be opportune to mention that large objectives do not transmit light proportionately with their increased diameter, owing to greater thickness of the lenses, which increases the absorption. The Washington objective of 25·8-inch aperture is 2·87 inches in thickness, and more than half the light which falls upon it is lost by absorption. On the other hand, specula, with every enlargement of aperture, give proportionately more light-grasping power, and their diameters might be greatly increased but for the mechanical obstacles in the way of their construction. Mr. Ranyard expresses the opinion that "with the refractor we are fast approaching the practical limit of size." After referring to the Washington object-glass as above, he says:— "If we double the thickness, more than three quarters of the light would be absorbed and less than one quarter would be transmitted. The greatest loss of light is only for the centre of the object-glass; but in all parts the absorption is qua- drupled for a lens of double aperture." If, therefore, future years see any great development in the sizes of telescopes, it will probably be in connection with reflectors; for the loss of light by absorption in the thick lenses of large refractors must ultimately determine their limits. Mr. Calver says:— "The light of reflectors exceeding 18 inches in diameter is certainly greater than that of refractors of equal size, and for anything like 3 feet very much greater." He nearly obtained the order for a monster reflector for the Lick Observatory, the Americans admitting that the reflector must be the instru- ment of the future for power and light because there were practically no limits to its size. But the reflector has not been much used in America, and therefore is little known. For this reason the authorities decided to erect a large refractor, and they appear to have been justified in their selection, for the 36-inch objective has proved excellent.

CHAPTER III.

NOTES ON TELESCOPES AND THEIR ACCESSORIES.

Choice of Telescopes.—The subject of the choice of telescopes has exercised every astronomer more or less, and the question as to the best form of instrument is one which has occasioned endless controversy. The decision is an important one to amateurs, who at the outset of their observing careers require the most efficient instruments obtainable at reasonable cost. It is useless applying to scientific friends who, influenced by different tastes, will give an amount of contradictory advice that will be very perplexing. Some invariably recommend a small refractor and unjustly disparage reflectors, as not only unfitted for very delicate work, but as constantly needing re-adjustment and resilvering *.

Others will advise a moderate-sized reflector as affording wonderfully fine views of the Moon and planets. The question of cost is greatly in favour of the latter construction, and, all things considered, it may claim an unquestionable advantage. A man who has decided to spend a small sum for the purpose not merely of gratifying his curiosity but of doing really serviceable work, must adopt the reflector, because refractors of, say, 5 inches and upwards are far too costly, and become enormously expensive as the diameter increases. This is not the case with reflectors ; they come

---

* My 10-inch reflector by With-Browning was persistently used for four years without being resilvered or once getting out of adjustment.

within the reach of all, and may indeed be constructed by the observer himself with a little patience and ingenuity.

Refractors and Reflectors.—The relative merits of refractors and reflectors * have been so frequently compared and discussed that we have no desire to re-open the question here. These comparisons have been rarely free from bias, or sufficiently complete to afford really conclusive evidence either way. There is no doubt that each form of instrument possesses its special advantages: aperture for aperture the refractor is acknowledged to be superior in light-grasping power, but the ratio given by different observers is not quite concordant. A silver-on-glass mirror of 8-inches aperture is certainly equal to a 7-inch objective in this respect, while as regards dividing power and the definition of planetary markings, the reflector is equal to a refractor of the same aperture. The much shorter focal length of the reflector is an advantage not to be overlooked. A century ago Sir W. Herschel figured his specula to foci of more than a foot to every inch of aperture, except in the case of his largest instruments. Thus he made specula of 18½-inches and 24-inches diameter, the former of which had a focal length of 20 feet and the latter of 25 feet. The glass mirrors of the present time are much shorter, and the change has not proved incompatible with excellent performance. Calver has made two good mirrors of 17¼-inches aperture, and only 8 ft. 4 in. focus. Mr. Common's 5-foot mirror is only 27½ feet, so that in these instances the length of the tube is less than six times the diameter.

It has long been proved that refractors and reflectors alike are, in good hands, capable of producing equally good results ; and we may depend upon it that, in spite of all argument and experiment, both kinds of telescope will continue to hold their own until superseded by a new combination, which hardly seems likely. If the observer is free from prejudice, he will have no cause to deplore the character of his instrument, always supposing it to be by a good maker. Be it

* In this and future references to reflectors the Newtonian form is alluded to. The direct-vision reflectors of Gregory and Cassegrain have gone out of use, and the present popularity of Newtonians may be regarded as a case of the " survival of the fittest."

object-glass or speculum, he will rarely find it lacking in effectiveness. It happens only too often that the telescope or

Fig. 12.

" The Popular Reflector " by Calver.

the atmosphere is hastily blamed when the fault rests with the observer himself. Let him be persistent in waiting oppor- tunities, and let the instrument be nicely adjusted and in good condition, and in the great majority of cases it will perform all that can reasonably be expected of it.

In choosing appliances for observational purposes, the observer will of course be guided by his means and requirements. If his inclination lead him to enter a particular department of research, he will take care to provide himself with such instruments as are specially applicable to the work in hand. Modern opticians have effected so many improvements, and brought out so many special aids to smooth the way of an observer, that it matters little in which direction he advances ; he will scarcely find his progress impeded by want of suitable apparatus. In size, as also in character, the observer should be careful to discriminate as to what is really essential. Large instruments and high powers are not necessary to show what can be sufficiently well seen in a small telescope with moderate power. Of course there is nothing like experience in such matters, and practice soon renders one more or less proficient in applying the best available means.

An amateur who really wants a competent instrument and has to consider cost, will do well to purchase a Newtonian reflector. A 4½-inch refractor will cost about as much as a

Fig. 13.

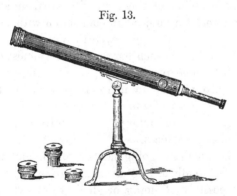

3-inch Refracting-Telescope, by Newton & Co.

10-inch reflector, but, as a working tool, the latter will possess a great advantage. A small refractor, if a good one, will do wonders, and is a very handy appliance, but it will not have sufficient grasp of light for it to be thoroughly serviceable on faint objects. Anyone who is hesitating in his choice should look at the cluster about χ Persei through instruments such as alluded to, and he will be astonished at the vast difference in favour of the reflector. For viewing sun-spots and certain lunar objects small refractors are very effective, and star-images are usually better seen than in reflectors, but the latter are much preferable for general work on account of three important advantages, viz., cheapness, illuminating power, and convenience of observation. When high magnifiers are employed on a refractor of small aperture, the images of planets become very faint and dusky, so that details are lost.

Observer's Aims.—If the intending observer merely requires a telescope to exhibit glimpses of the wonders which he has seen portrayed in books, and has no intention of pursuing the subject further than as an occasional hobby, he will do well to purchase a small refractor between 3 and 4 inches in aperture. Such instruments are extremely effective on the Sun and Moon, which are naturally the chief objects to attract attention, and, apart from this, appliances of the size alluded to may be conveniently used from an open window. The latter is an important consideration to many persons; moreover, a small telescope of this kind will reveal an astonishing number of interesting objects in connection with the planets, comets, &c., and it may be employed by way of diversion upon terrestrial landscape, as such instruments are almost invariably provided with non-inverting eyepieces. Out-of-door ob_serving is inconvenient in many respects, and those who procure a telescope merely to find a little recreation will soon acknowledge a small refractor to be eminently adapted to their purposes and conveniences.

Those who meditate going farther afield, and taking up observation habitually as a means of acquiring practical knowledge, and possibly of doing original work, will essentially need different means. They will require reflectors of about 8 or 10 inches aperture; and, if mounted in the open on solid ground, so much the better, as there will be a more expansive view, and a freedom from heated currents, which renders an apartment unsuited to observations, unless with small apertures where the effects are scarcely appreciable. A reflector of the diameter mentioned will command sufficient grasp to exhibit the more delicate features of planetary markings, and will show many other difficult objects in which the sky abounds. If the observer be specially interested in the surface configuration of Mars and Jupiter he will find a reflector a remarkably efficient instrument. On the Moon and planets it is admitted that its performance is, if not superior, equal to that of refractors. If, however, the inclination of the observer leads him in the direction of double stars, their discovery and measurement, he will perhaps find a refractor more to be depended upon, though there is no reason why a well-mounted reflector should not be successfully employed in this branch ; and the cost of a refractor of the size to be really useful as an instrument of discovery must be something very considerable—perhaps ten times as great as that of a reflector of equal capacity. As far as my own experience goes the refractor gives decidedly the best image of a star. In the reflector, a bright star under moderately high power is seen with rays extending right across the field, and these appear to be caused by the supports of the flat.

Testing Telescopes.—No amateur should buy an instrument, especially a second-hand one, without testing it, and this is a delicate process involving many points to be duly weighed. Experience is of great service in such matters, and is, in fact, absolutely necessary. Even old observers are sometimes misled as to the real worth of a glass. In such cases, there is nothing like having a reliable means of comparison, *i. e.* another telescope of acknowledged excellence with which to test the doubtful instrument. In the absence of such a standard judgment will be more difficult, but with care a satisfactory decision may be arrived at. The Moon is too easy an object for the purpose of such trials ; the observer should rather select Venus or Jupiter. The former is, however, so brilliant on a dark sky, and so much affected with glare, that the image will almost sure to be faulty even if the glass is a good one. Let the hour be either near sunrise or sunset, and if the planet has a tolerably high altitude, her disk ought to be seen beautifully sharp and white. Various powers should be tried, increasing them each time, and it should be

noticed particularly whether the greater expansion of the
image ruins the definition or simply enfeebles the light.   In
a thoroughly good glass faintness will come on without seriously impairing the definite contour of the object viewed,
and the observer will realize that the indistinctness is merely
occasioned by the power being relatively in excess of the
light-grasp.   But in a defective telescope, a press of magnifying power at once brings out a mistiness and confuses the
details of the image in a very palpable manner.   Try how he
will, the observer will find it impossible to get rid of this,
except, perhaps, by a " stop " which cuts off so much light
that the instrument is ineffective for the work required of it.
The blurred image is thought, at the moment of its first perception, to be caused by the object being out of focus, and
the observer vainly endeavours to get a sharper image until
he finds the source of error lies elsewhere.   A well-figured
glass ought to come very sharply to a focus.   The slightest
turn of the adjusting-screw should make a sensible difference.
On the other hand, an inferior lens will permit a slight alteration of focusing without affecting the distinctness, because
the rays from the image are not accurately thrown to a point.
Jupiter is also a good test.   The limbs of the planet, if shown
clean and hard, and the belts, if they are pictured like the
finely cut details of an engraving, will at once stamp a telescope as one of superior quality.   Saturn can also be examined
though not, perhaps, so severe a test.   The belts, crape ring,
Cassini's division, ought to be revealed in any telescope of
moderate aperture.   If, with regard to any of these objects,
the details apparently run into each other and there is a "fuzzy"
or woolly aspect about them which cannot be eliminated by
careful focusing, then either the atmosphere or the telescope
is in fault.   If the former, another opportunity must be
awaited.   An observer of experience will see at a glance
whether the cause lies in the air or the instrument.   The
images will be agitated by obnoxious currents, if the defects
are due to the atmosphere, but if the glass itself is in error,
then the objects will be comparatively tranquil but merged
in hazy outlines, and a general lack of distinctness will be
apparent.   Perhaps the best test of all as to the efficiency of

a telescope is that of a moderately bright star, say of the 2nd or 3rd magnitude. With a high power the image should be very small, circular, and surrounded by two or three rings of light lying perfectly concentric with each other. No rays, wings, or extraneous appearance other than the diffraction rings should appear.

This, however, specially applies to refractors, for in reflectors the arms of the flat occasion rays from any bright star ; I have also seen them from Mars, but of course this does not indicate an imperfect mirror. If there is any distortion on one side of the image, then the lenses are inaccurately centred though the instrument may be otherwise good, and a little attention may soon set matters right. When testing a glass the observer should choose objects at fairly high altitudes, and not condemn a telescope from a single night's work unless the evidence is of unusually convincing character. If false colour is seen in a silver-on-glass reflector it is originated by the eyepiece, though not necessarily so in a refractor. The object-glass of the latter will be sure to show some uncorrected colour fringing a bright object. A good lens, when exactly focused, exhibits a claret tint, but within the focus purple is seen and beyond the focus green comes out. In certain cases the secondary spectrum of an object-glass is so inadequately corrected that the vivid colouring of the images is sometimes attributed by inexperienced observers to a real effect. A friend who used a 3-inch refractor once called on me to have a glimpse of Jupiter through my 10-inch With-reflector. On looking at the planet he at once exclaimed "But where are the beautiful colours, Mr. Denning ? " I replied to his question by asking another, viz., " What colours ? " he answered, " Why, the bright colours I see round Jupiter in my refractor ? " I said, " Oh, they exist in your telescope only ! " He looked incredulous, and when he left me that night did not seem altogether pleased with the appearance of Jupiter shorn of his false hues !

Mounting.—Too much care cannot be given to the mounting of telescopes, for the most perfectly figured glass will be rendered useless by an inefficient stand ; a faulty lens, if thoroughly well mounted, will do more than a really good one on a shaky or unmanageable mounting. Whatever form is adopted, the arrangement should ensure the utmost steadiness, combined with every facility for readily following objects. A man who has every now and then to undergo a great physical exertion in bodily shifting the instrument is rendered unfit for delicate work. The telescope should be provided with every requisite for carrying on prolonged work with slight exertion on the part of the observer. Unless the stand is firm there will be persistent vibrations, especially if the instrument is erected in the open, for there are very few nights in the year when the air is quite calm. These contingencies should be provided against with scrupulous attention if the observer would render his telescope most effective for the display of its powers, and avoid the constant annoyance that must otherwise follow.

Eyepieces.— Good eyepieces are absolutely essential. Many object-glasses and specula have been deprecated for errors really originated by the eyepiece. Again, telescopes have not unfrequently been blamed for failures through want of discrimination in applying suitable powers. A consistent adaptation of powers according to the aperture of the telescope, the character of the object, the nature of the observation, and the atmospheric conditions prevailing at the time, is necessary to ensure the best results. If it is required to exhibit a general view of Jupiter and his satellites to a friend, we must utilize a low power with a large field ; if, on the other hand, we desire to show the red spot and its configura-

Fig. 14.

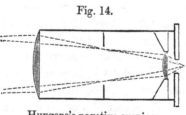

Huygens's negative eyepiece.

tion in detail, we must apply the highest power that is satisfactorily available. The *negative* or Huygenian eyepiece is the one commonly used, and it forms good colourless images, though the field is rather small. The *positive* or Ramsden eyepiece gives a flatter and larger field, but it is not often achromatic. A Kellner eyepiece, the feature of which is a very large field, is often serviceable in observations of nebulæ, clusters, and comets. Telescopes are sometimes stated to bear 100 to the inch on planets, but this is far beyond their

Fig. 15.

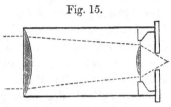

Ramsden's positive eyepiece.

capacities even in the very best condition of air. Amateurs soon find from experience that it is best to employ those powers which afford the clearest and most comprehensive views of the particular objects under scrutiny. Of course when abnormally high powers are mentioned in connection with an observation, they have an impressive sound, but this is all, for they are practically useless for ordinary work. I find that 40, or at the utmost 50 to the inch, is ample, and generally beyond the capacities of my 10-inch reflector. A Barlow lens used in front of the eyepiece raises the power about one third, and thus a whole set of eyepieces may be increased by its insertion. It is said to improve the definition, while the loss of light is very trifling. I formerly used a Barlow lens in all planetary observations, but finally dispensed with it, as I concluded the improved distinctness did not compensate for the fainter image. A great advantage, both in light and definition, results from the employment of a single lens as eyepiece. True, the field is very limited, and, owing to the spherical aberration, the object so greatly distorted near the edges that it must be kept near the centre, but, on the whole, the superiority is most evident. By many careful trials I find it possible to glimpse far more detail in planetary markings than with the ordinary eyepiece. Dawes, and other able observers, also found a great advantage in the single lens, and Sir W. Herschel, more than a century ago, expressed himself thus :—" I have tried both the double and single lens eye-glass of equal powers, and always found that the single eye-glass had much the superiority in light and distinctness."

Requisite Powers.—For general purposes I believe three eyepieces are all that is absolutely requisite, viz., a low power with large field for sweeping up nebulæ and comets ; a moderate power for viewing the Moon and planets ; and a high power for double stars and the more delicate forms on the planets. For a 3-inch refractor, eyepieces of about 15, 75, and 150 would be best, and for a 10-inch reflector 40, 150, and 300. For very difficult double stars a still higher power will be occasionally useful, say 250 for the refractor, and 500 for the reflector. The definition usually suffers so much under high powers, and the tremors of the atmosphere are brought out so conspicuously, that the greater expansion of the image of a planet does not necessarily enable it to present more observable detail. The features appear diluted and merged in hazy outlines, and there is a lack of the bright, sharply determinate forms so steadily recognized under lower magnifiers. In special cases great power may become essential, and under certain favourable circumstances, will prove really serviceable, but, in a general way, it is admitted that the lowest power which shows an object well is always the best. I have occasionally obtained very fair views of Saturn with a power of 865, but find that I can perceive more of the detail with 252. Some daylight observations of Venus were also effected under very high power, and, though the definition remained tolerably good, I found as the result of careful comparison that less power answered more satisfactorily. But it would be absurd to lay down inviolable rules in such cases. Special instruments, objects, and circumstances require special powers, and observers may always determine with a little care and experience the most eligible means to support their endeavours. One thing should be particularly remembered, that the power used must not be beyond the illuminating capacity of the instrument, for planetary features appear so faint and shady under excessive magnifiers that nothing is gained. To grasp details there must be a fair amount of light. I have seen more with 252 on my 10-inch reflector than with 350 on a 5¼-inch refractor, because of the advantage from the brighter image in the former case.

Overstating Powers.—It seems to be a fashionable imposition on the part of opticians to overstate magnifying powers. Eyepieces are usually advertised at double their true strength. My own 10-inch reflector was catalogued as having four eyepieces, 100 to 600, but on trial I found the highest was no more than 330. This custom of exaggerating powers seems to have long been a privileged deception, and persons buying telescopes ought to be guarded against it. Dr. Kitchiner says it originated with the celebrated maker of reflectors, James Short, and justly condemns it as a practice which should be discontinued. I suppose it is thought that high powers advertised in connection with a telescope have an exalted sound and are calculated to attract the unwary purchaser ; but good instruments need no insidious trade artifices to make them saleable. The practice does not affect observers of experience, because it is well understood, and they take good care to test their eyepieces directly they get them. But the case is different with young and inexperienced amateurs, who naturally enough accept the words of respectable opticians, only to find, in many cases, that they have been misleading and a source of considerable annoyance.

Method of finding the Power.—The magnifying power of a telescope may be determined by dividing the focal length of the object-glass or mirror by the focal length of the eye-lens. Thus, if the large glass has a focus of 70 inches and the eye-lens a focus of one inch, then the power is 70. If the latter is only ¼-inch focus, the resulting power will be 280. But this method is only applicable to single lens eyepieces. We may, however, resort to several other means of finding the powers of the compound eyepieces of Huygens or Ramsden. Let the observer fix a slip of white cardboard, say 1 inch wide, to a door or post some distance off, and then (with a refractor) view it, while keeping the disengaged eye open, and note the exact space covered by the telescopic image of the card as projected on the door seen by the other eye. The

E

number of inches included in the space alluded to will represent the linear magnifying power. A brick wall or any surface with distinct, regularly marked divisions will answer the same purpose, the number of bricks or divisions covered by the telescopic image of one of them being equivalent to the power. But it should not be forgotten that a telescope magnifies slightly less upon a celestial object than upon a near terrestrial one owing to the shorter focus, and a trifling allowance will have to be made for this. Another plan may be mentioned. When the telescope is directed to any fairly bright object or to the sky, and the observer removes his eye about 10 inches from the eyepiece, a sharply defined, bright little disk will be perceived in the eye-lens. If the diameter of this disk is ascertained and the clear aperture of the object-glass or mirror is divided by it, the quotient will be the magnifying power. Thus, if the small circle of light is ·2 inch diameter and the effective aperture of the large glass 5 inches, then the power is 25. If the former is ·02 inch diameter and the latter 7·5 inches, the power will be 375. The dynamometer is a little instrument specially designed to facilitate this means of fixing the magnifying power. It enables the diameter of the small luminous circle in the eye-lens to be very accurately measured, and this is a most important factor in deriving the power by this method.

Fig. 16.

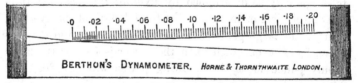

BERTHON'S DYNAMOMETER. *HORNE & THORNTHWAITE LONDON.*

Field of Eyepiece.—Observers often require to know the diameter of the fields of their eyepieces. Those engaged in sweeping up comets, nebulæ, or other objects requiring large fields and low powers, find it quite important to have this information. They may acquire it for themselves by simple methods. A planet, or star such as δ Orionis, η or γ Virginis, or η Aquilæ, close to the equator, should be allowed to run

exactly through the centre of the field, and the interval
occupied in its complete transit from ingress to egress noted
several times. The mean result in min. and sec. of time
must then be multiplied by 15, and this will represent the
diameter required in min. and sec. of arc on the equator.
A planet or star near the meridian is the best for the purpose.
If the object occupies 1 min. 27 sec. of time in passing from
the E. to the W. limit of the field, then 87 sec. × 15 = 1305″,
or 21′ 45″. A more accurate method of deriving the angle
subtended by the field is to let a star, say Regulus, pass
through the centre, and fix the time which lapses in its entire
passage by a sidereal clock ; then the interval so found
× 15 × cosine of the declination of Regulus will indicate the
diameter of the field. Suppose for instance, that the star
named occupies 2 min. 14 sec. = 134 sec. in its passage right
across the whole and central part of the field : then

$$
\begin{array}{lr}
134 \log & 2\!\cdot\!127105 \\
15 \log & 1\!\cdot\!176091 \\
\text{Dec. of Regulus } 12°\ 30'\ \log \cos & 9\!\cdot\!989581 \\
\hline
1962''\ \log & 3\!\cdot\!292777
\end{array}
$$

so that the diameter of the field of the eyepiece must be 32′
42″, nearly corresponding with the diameter of the Moon.

Limited Means no Obstacle.—There are many observers who,
having limited means, are apt to consider themselves practically unable to effect good work. This is a great illusion.
There are several branches of astronomy in which the diligent
use of a small instrument may be turned to excellent account.
Perseverance will often compensate for lack of powerful
appliances. Many of the large and expensive telescopes, now
becoming so common, are engaged in work which could be as
well performed with smaller aperture, and when the manifold
advantages of moderate instruments are considered, amateurs
may well cease to deplore the apparent insufficiency of their
apparatus. It is, however, true that refractors have now
attained dimensions and a degree of proficiency never contemplated in former times, and that the modern ingenuity of art
has given birth to innumerable devices to facilitate the work
of those engaged in observation. In many of our best

Fig. 17.

Cooke and Son's Educational Telescope.

appointed observatories the arrangements are so very replete with conveniences, and so sedatory in their influences, that the observer has every inducement to fall asleep, though we do not find instances of " nodding " recorded in their annals. Further progress in the same direction leads us to joyfully anticipate the time when, instead of standing out in the frost, we may comfortably make our observations in bed. This will admirably suit all those who, like Bristol people, are re- ported to sleep with one eye open! But, to be more serious, the work of amateurs is much hindered by lack of means to construct observatories wherein they may conduct researches without suffering from all the rigours of an unfavourable climate. Many of them have, like William Herschel a century ago, to pursue their labours under no canopy but the heavens above, and are exposed to all the trying severity of frost and keen winds, which keep them shivering for hours together, and very much awake !

Observing-Seats.—As to observing-seats, many useful con- trivances have been described from time to time in the Astronomical Register ' and ' English Mechanic.' Some of these answer their design admirably, but I believe a good chair, embodying all the many little requirements of the observer, yet awaits construction. Those I have seen, while supplying certain acknowledged wants, are yet deficient in some points which need provision. With my reflector I find an ordinary step-ladder answers the purpose very well. It is at once light, simple, and durable, and enables observations to be secured at any altitude. It may be readily placed so that the observer can work in a sitting posture, and the upper shelves, while convenient to lean upon, may be so arranged as to hold eyepieces, and are to be further utilized when making draw- ings at the telescope. I find it possible to obtain very steady views of celestial objects in this way. Everyone knows that during a critical observation it is as essential for the observer to be perfectly still as it is for the instrument to be free from vibration. A person who stands looking through a telescope feels a desire to ensure a convenient stability by catching hold of it. The impression is no doubt correctly conveyed to his mind that he may obtain a better view in this way ; and so he would, were it not for the dancing of the image which instantly follows the handling of the instrument. For this reason it is absolutely necessary that no part of the observer touch the telescope while in use. He must ensure the desired steadiness, which is really a most important consideration, by other means ; and an observer who provides for this contingency will have taken a useful step in the way of achieving delicate work.

Advantage of Equatoreals.—Those who employ equatoreal mounting and clock-work will manifestly command an advantage in tracing features on a planet or other object requiring critical scrutiny. Common stands, though often good make-shifts, require constant application on the part of the observer, when his undivided attention should be concentrated on the object. With an alt-azimuth stand nearly one half the observer's time is occupied in keeping the object near the middle of the field. Though good views are obtainable, they are very fugitive. Just as the delicate features are being impressed on the retina they are lost in the ill-defined margin of the field, or from the necessity of suddenly shifting the object back. A succession of hurried views of this kind, during which the observer is frantically endeavouring to grasp details which only require a steady view to be well displayed, are often tantalizing and seldom satisfactory in their issues. This is especially the case when a single lens and high powers are used, and if the night is windy the difficulty is intensified. It is, therefore, evident that a clock-driven telescope possesses marked advantages in delicate work on faint objects, because the prolonged view better enables the eye to gather in the details which are all but lost in the elusive glimpses afforded by inferior means. Still we must not forget that rough appliances do not present an effectual barrier to success. The very finest definition comes only in momentary glimpses. The sharply-cut outlines of planetary configuration cannot steadily be held for long together. Only now and then the image acquires the distinctness of an engraving, when the air and the focus of eye and telescope severally combine to produce a perfect picture. Observers, therefore, whose instruments are simply, though perhaps substantially mounted in handy fashion, must profit by these moments of fine seeing, and, when drawing, will find it expedient to fill in, little by little, the delicate forms which reach the eye. This will take much time owing to the drawbacks alluded to, but the outcome will more than justify its expenditure, and the observer will gain patience and perseverance which will prove a useful experience in the future.

Lenses out of centre or misplaced are, like other defects, calculated to give rise to errors as numerous as they are various. But the most striking of these apparently belong to a period when telescopes were far less perfect and popular than at the present day. Indeed, it is surprising that so very few false or imaginary discoveries are announced when we consider the vast array of instruments that are now employed. It is true we occasionally hear that a comet has been discovered close to Jupiter, that several companions have been seen to Polaris, or that some other extraordinary " find " has been effected, but the age is dead when such announcements were accepted without suitable investigation. The satellite of Venus has long since ceased to exist. The active volcanoes on the Moon have become extinct. Even Vulcan will have to be set aside, and, like many another sensation which caused quite a *furóre* in its day, must soon be altogether expunged from the category of " suspects."

Test-objects. Opticians sometimes advertise lists of objects— generally double stars—which may be seen with their instruments, but it does not appear to be sufficiently understood that the character of a telescope is dependent in a great degree upon the ability of the observer, who can either make or mar it, according to the skill he displays in its management. Some men will undoubtedly see more with 5 inches of aperture than others will with 10. Certain observers appear to excel in detecting delicate planetary markings, while others possess special aptitude for glimpsing minute objects such as faint satellites, or *comites* to double stars, and the explanation seems to be that partly by experience and partly by differences in the sensitiveness of vision, exceptional powers are sometimes acquired in each of these departments.

The various test-objects which have been given by reliable
authorities, though representing average attainments, are not
applicable to the abnormal powers of vision possessed by
certain observers. In fact, the capacity of a telescope cannot
be correctly assigned and its powers circumscribed by ar-
bitrary rules, because, as already stated, the character of the
observer himself becomes a most important factor in this
relation. Climatic influences have also considerable weight,
though less so than the personal variations referred to, for
one man will succeed, where another meets with utter failure.
This is unquestionably due to differences in eyesight, method,
and experience. But whatever the primary causes may be,
everyone knows they induce widely discordant results, and
occasion many of the contradictions which become the subjects
of controversy. And, as a rule, amateurs should avoid controversy, because it very rarely clears up a contested point.
There is argument and reiteration, but no mutual understanding or settlement of the question at issue. It wastes
time, and often destroys that good feeling which should subsist amongst astronomers of every class and nationality. In
cases where an important principle is involved, and discussion
promises to throw light upon it, the circumstances are quite
different. But paltry quibblings, fault-finding, or the constant
expression of negative views, peculiar to sceptics, should be
abandoned, as hindering rather than accelerating the progress
of science. Let observers continually exercise care and
discretion and satisfy themselves in every legitimate way as
to the accuracy of their results, and they may fearlessly give
them expression and overcome any objections made to their
acceptance. They should accord one another an equal desire
for the promotion of truth. Competition and rivalry in good
spirit increase enthusiasm, but there is little occasion for the
bitterness and spleen sometimes exhibited in scientific journals.
There are some men whose reputations do not rest upon good
or original work performed by themselves, but rather upon
the alacrity with which they discover grievances and upon
the care they will bestow in exposing trifling errors in the
writings of their not-infallible contemporaries. Such critics
would earn a more honourable title to regard were they to

devote their time to some better method of serving the cause of science.

Cheapness and increasing number of Telescopes.—A marked feature of optical instruments is their increasing cheapness. Little more than half a century ago Tulley charged £315 for a 10-inch Newtonian reflector. At the present time Calver asks £50 for an instrument of the same aperture, and sometimes one may be picked up, second-hand, for half of that amount. Not only have telescopes become cheaper, but they have greatly improved in performance since silvered glass superseded the metallic speculum. Hence we find moderately-powerful instruments in the hands of a very large number of observers. Astronomical publications have proportionately increased, so that amateurs of to-day can boast of facilities, both of making and recording observations, which were scarcely dreamt of a century ago. It must be admitted, however, that the results hardly do justice to the means available. Such an enormous number of telescopes are variously employed that one cannot avoid a feeling of surprise at the comparative rarity of new discoveries, and, indeed, of published observations generally. It is certain that the majority of existing telescopes are either lying idle or applied in such a desultory fashion as to virtually negative the value of the results. Others, again, are indiscriminately employed upon every diversity of object without special aim or method, and with a mere desire to satisfy curiosity. Now it is to be greatly deplored that so much observing strength is either latent or misdirected. The circumstances obviously demand that an earnest effort should be made to utilize and attract it into suitable channels. To do this effectually, the value of collective effort should be forcibly explained, the interest and enthusiasm of observers must be aroused in a permanent manner, and they must be banded together according to their choice of subjects. An effort in this direction has been made by the Liverpool Astronomical Society, and the results have proved distinctly favourable ; a considerable amount of useful work has been effected in several branches and it forms the subject of some valuable reports which have been annually published in the 'Journal.'

Utility of Stops.—There are a good many details connected
with observation which, though advice may be tendered in a
general way, are best left to the discrimination of observers,
who will very soon discover their influences by practical
trial and treat them accordingly.   The employment of stops
or diaphragms to contract the aperture of telescopes is a
question on which a diversity of opinion has been expressed.
It is often found, on nights of indifferent seeing, that the
whole aperture, especially of a faulty instrument, gives bad
images, and that, by reducing it, definition becomes immensely improved.   But Mr. Burnham, the double star
observer, records his opinion that a good glass needs no contraction, and that the whole aperture shows more than a part
unless there is defective figuring at the outer zone of the
lens, which will be cut off by the stop and its performance
thereby greatly improved.   He seems to think that a glass
requiring contraction is essentially defective, but this is
totally opposed to the conclusions of other observers.   It is
almost universally admitted that, on bad nights, the advantages of a large aperture are neutralized by unsteady definition, and that, by reducing the diameter, the character
of the images is enhanced.   As regards instruments of
moderate calibre the necessity is less urgent.   With my
10-inch reflector I rarely, if ever, employ stops, for by
reducing the aperture to 8 inches the gain in definition does
not sufficiently repay for the serious loss of light.   But in the
case of large telescopes the conservation of light is not so
important, and a 14-inch or 16-inch stop may be frequently
employed on an 18-inch glass with striking advantage.   The
theory that only defective lenses improve with contraction is
fallacious, for in certain cases where stops are regularly
employed it is found that, under circumstances of really good
seeing, the whole aperture gives images which are as nearly
perfect as possible.   It is clear from this that the fault lies
with the atmosphere, and that under bad conditions it
becomes imperative to limit its interference consistently with
the retention of sufficient light to distinguish the object well.
In large reflectors, particularly, the undulations of the air
are very active in destroying definition, and the fact will be

patent enough to anyone who compares the images given in widely different apertures. The hard, cleanly cut disks shown by a small speculum or object-glass offer an attractive contrast to the flaring, indefinite forms often seen in big telescopes.

Cleaning Lenses.—As to wiping objectives or mirrors, this should be performed not more often than absolute necessity requires ; and in any case the touches should be delicate and made with materials of very soft texture. The owner of a good objective should never take the handkerchief out of his pocket and, in order to remove a little dust or dew, rub the glass until the offensive deposit is thought to be removed. Yet this is sometimes done, though frequent repetition of such a process must ultimately ruin the best telescope notwith-standing the hardness of the crown glass forming the outer lens of the objective. It will not bear such "rough and ready" usage and in time must show some ugly scratches which will greatly affect its value though they may not seriously detract from its practical utility. Good tools deserve better treatment. When the glass really wants cleaning, remove it from the tube and sweep its whole surface gently with a dry camel's-hair brush, or when this is not at hand get a piece of linen and "flick" off the dust particles. Then wipe the lens, as soon as these have been dislodged, with an old silk, or soft cambric handkerchief ; fine chamois leather is also a good material, and soft tissue paper, aided by the breath, has been recommended. But whatever sub-stance may be adopted it must be perfectly clean and free from dust. When not in use it should be corked up in a wide-necked bottle where it will be safe from contact with foreign particles. In the case of mirrors there is an obvious need that, when being repolished, the material used should be perfectly dry and that the mirror also should be in the same state. It is unnecessary to say here that in no case must the silver film be touched when it is clouded over with moisture. This must first be allowed to evaporate in a free current of air or before a fire ; the former is to be preferred. A suitable polishing-pad may be made with a square piece of washleather or chamois in which cotton-wool is placed and then tied into a bag. This may be dipped into a little of the finest rouge, and its employment will often restore a bright surface to the mirror. But the latter should be left " severely alone " unless there is urgent occasion to repolish it, as every application of the rouged pad wears the film and may take off minute parts of it, especially when dust has not been altogether excluded. The precarious nature of the silvered surface undoubtedly constitutes the greatest disadvantage of modern reflectors. The polish on the old metallic mirrors was far more durable. Some of Short's, figured 150 years ago, still exist and are apparently as bright as when they were turned out of the workshop ! I have a 4-inch Gregorian by Watson which must be quite a century old, and both large and small specula seem to have retained their pristine condition.

With regard to the duration of the silver-on-glass films, much of course depends upon the care and means taken to preserve them. Calver says that sometimes the deposit does not last so long as expected, though he has known the same films in use for ten years. A mirror that looks badly tarnished and fit for nothing will often perform wonderfully well. With my 10-inch in a sadly deteriorated state I have obtained views of the Moon, Venus, and Jupiter that could hardly be surpassed. The moderate reflection from a tarnished mirror evidently improves the image of a bright object by eliminating the glare and allowing the fainter details to be readily seen. When not in use a tight-fitting cap should always be placed over the mirror, and if a pad of cotton wadding of the same diameter is made to inlay this cap it tends to preserve the film by absorbing much of the moisture that otherwise condenses on its surface. The ' Hints on Reflecting-Telescopes,' by W. H. Thornthwaite and by G. Calver, and the ' Plea for Reflectors,' by J. Browning, may be instructively consulted by all those who use this form of instrument. The latter work is now, however, out of print, and Mr. Browning tells me that he has quite relinquished the manufacture of reflecting-telescopes. Mr. G. With of Hereford, who formerly supplied the mirrors for his instruments, has recently disposed of his reserve stock and entered an entirely different sphere of labour. In the publications above alluded to amateurs will find a large amount of practical information on the value and treatment of glass mirrors.

Opera-Glass.—A very useful adjunct, and often a really valuable one to the astronomical amateur, is the Opera-Glass, or rather the larger form of this instrument generally known as the Field-Glass. Of certain objects it gives views which cannot be surpassed, and it is especially useful in observations of variable stars and large comets. Whenever the horizon is being scanned for a glimpse of the fugitive Mercury, or when it is desired to have a very early peep at the narrow crescent of the young Moon, or to pick up Venus at midday, or Jupiter before sunset, all one has to do is to sweep over the region where the object is situated, when it is pretty sure to be caught, and the unaided eye will probably reach it soon afterwards. The opera-glass has the dignity of being the first telescope invented, for even its binocular form is not new; it is virtually the same pattern of instrument that was introduced at Middleburg in 1609, though its compound object-glasses are of more modern date. Anyone who entertains any doubts as to the efficacy of the opera-glass or has had little experience in its use will do well to look at the Pleiades and compare the splendid aspect of that cluster, as it is there presented, with the view obtained by the naked eye, and he will acknowledge at once that it constitutes a tool without which the observer's equipment is by no means perfect. The object-glasses should have diameters of 2 or 2½ inches, and the magnifying power lie between 4 and 6. There is a large field of view and the images are very bright. The observer is enabled to enjoy the luxury of using both his eyes, and when he directs the instrument upon a terrestrial landscape he will be gratified that it does not turn the world upside down! It is not surprising that an appliance, with recommendations so significant, is coming more into favour every day, and for those branches suitable to its means it is doing much useful work. A volume has been recently published dealing expressly with the use of the opera-glass in Astronomy; and in the 'Journal of the L.A.S.' vol. vii. p. 120, there is an excellent paper by Major Markwick on the same subject. This instrument will never, of course, by the nature of its

construction, be comparable to a modern telescope in regard
to power, for Galilei, when he augmented his magnifiers to
30, appears to have practically exhausted the resources of
this appliance.   But in all those departments requiring an
expansive field and little power with a brilliant and distinct
image, the larger form of opera-glass is a great desideratum,
and its portability is not one of the least of its advantages.

Dewing of Mirrors.—The disposition of mirrors to become
clouded over upon rises of temperature is a point meriting
comment. When permanently left in a telescope, fully exposed
out of doors, the speculum undergoes daily transitions.   The
heat generated in the interior of the tube by the sun's action
causes a thick film of moisture to form upon the silvered
surface of the mirror, which remains in this state for a
considerable time, though the moisture evaporates before
the evening.   The flat is similarly affected, and the result
of these frequent changes is that the coating of silver
becomes impaired and presents a crackly appearance all over
the surface.   Sometimes when a marked increase of temperature occurs towards evening the speculum is rendered
totally unserviceable until it has been submitted to what
Dr. Kitchiner terms a process of "roasting."   The vapour
will soon disappear when the mirror is brought indoors and
placed before a fire ; but it is not till some time after it has
been remounted in the tube that it will perform satisfactorily.
Those who keep their mirrors in more equable temperatures
will not experience these inconveniences, which may also in
some measure be obviated by regularly placing a tight-fitting
cap, inlaid with cotton-wool, over the speculum at the conclusion of work.   This also protects the silver from the yellow
sulphurous deposit which soon collects upon it if used in a
town.   All sudden variations of temperature act prejudicially
on the performance of specula, and their best work is only
accomplished when free from such disturbing elements.   I
have rarely found the flat to become dewed in a natural
way during the progress of observation.   If on a cold
night the observer puts his hand upon its supports in order
to alter its adjustment it instantly becomes dewed, or if
he stands looking down the tube it is almost sure to be

similarly affected ; but in the ordinary course of work the flat is little liable to become dewed in sensible degree. With refractors dew-caps are very necessary, though they do not always prevent the deposition of moisture on the object-glass, and this occasions frequent wiping or drying, which in either case is very objectionable.

Celestial Globe.—This forms another extremely useful addendum to the appliances of the amateur. It enables a great many problems to be solved in a very simple manner, and helps the young student to a lucid comprehension of the apparent motions and positions of the fixed stars. With ' Keith on the Globes ' as a reference-book he may soon acquire the method of determining the times of rising, southing, and setting of any celestial object the place of which is known. He can also readily find the height (altitude) and bearing (azimuth) at any time. The distance in degrees between any two stars or between a star and the Moon, a planet, or a comet may be found at a glance by laying the quadrant of altitude on the pair of objects and reading off the number of degrees separating them. If a new comet has been discovered, its position should be marked in pencil upon the globe ; and the observer, after having noted its exact place relatively to neighbouring stars, may proceed to identify the object with his telescope. If a large meteor is seen, its apparent path amongst the constellations should be projected on the globe and the points, in R.A. and Dec., of beginning and ending of the flight read off and entered in a book. In many other practical branches of astronomy this instrument will prove highly serviceable, and is far preferable to a star-atlas. But the latter is the most useful to the beginner who is just learning the names of the stars and the configuration of the chief groups, because on the globe the positions are all reversed east and west. The surface of the globe represents the entire star-sphere reduced to a common distance from the earth, and as seen from outside that sphere. The observer, therefore, must imagine his eye to be situated in the centre of the globe, if he would see the stars in the same relative places as he sees them in the heavens. The reversion of the star-positions to which we have been alluding is very confusing at first, and no doubt it provokes mistakes, but a little experience will practically remove this objection. The one great recommendation to a star-atlas is that it displays the stars in the natural positions in which they are discerned by the eye, thus enabling the student to become readily acquainted with them, whereas the celestial globe affords no such facility. But in other respects the latter possesses some valuable functions, and the amateur who devotes some of his leisure to mastering the really useful problems will attain a knowledge that will be of great benefit to him in after years. A globe of 12-inches diameter will be large enough for many purposes, but one of 18-inches will be the most effective size. It should be mounted on a tall stand with single body and tripod base. The stands, fitted with three parallel legs, in which the globe is supported in the middle by weak connections from them, are not nearly so durable. I have used several 18-inch globes mounted in this manner, and the supports have quite given way under the pressure of constant use; but this is impossible with the strong single body, which is capable of withstanding any strain. Globes are frequently to be obtained second-hand, and at trifling cost; but the observer must allow for precession if he uses an old article. Many of the stars will be 1° or 2° east of the positions in which they are marked on the globe; and it will be necessary to remember this if the appliance is to be employed for exact results.

Observatories.—Massive and lofty buildings have long gone out of fashion, and lighter, drier structures have properly supplanted them. Instruments of size are generally placed on or near the ground and solidly supported to ensure stability, while the other erections are made consistent with the necessity for pretty equable temperature and freedom from damp. Amateurs will ordinarily find that a simple wooden enclosure for the telescope, with suitable arrangements for opening the top in any direction, is sufficient for their purpose and very inexpensive. Some observers have, indeed, secured the desired shelter for themselves and their telescopes by means of a canvas tent provided with ready means for obtaining sky-room. Berthon has given a good description of an amateur's observing-hut in 'The English Mechanic' for October 13th and 20th, 1871; and Chambers supplies some information about amateur observatories in 'Nature' for November 19th, 1885 *. Mr. Thornthwaite's 'Hints on Telescopes' may be usefully consulted for details of the Romsey Observatory, which, like the Berthon model, seems peculiarly adapted to the necessities of the amateur. The great requirements in such structures are that they should be dry and not obstruct any region of the firmament. They should also be large enough to allow the observer perfect freedom in his movements and during the progress of his observations. They are then decided advantages, and will materially add to that comfort and convenience without which it is rarely possible to accomplish really good work. When an observatory is to be dispensed with it becomes necessary to erect a small wooden house near the instrument, especially if placed at the far end of a garden, in which the observer may keep certain appliances, such as a lantern, celestial globe, step-ladder or observing-seat, oil, &c. Here also he may record his seeings, complete his sketches, and consult his working-list, star-charts, and ephemerides. A shelter of this sort, apart from its practical helpfulness, avoids any necessity for the observer to go in and out of doors, up and down stairs, &c., to the annoyance of the rest of his family, who, on a frosty night, are decidedly not of an astronomic turn, and vastly prefer house-warming to stargazing!

---

* Chambers's 'Descriptive Astronomy,' 4th ed. vol. ii., also contains some useful references and diagrams.

APTER IV.

NOTES ON TELESCOPIC WORK.

Preparation.—Working-Lists.—Wind.—Vision.—Records.—Drawing.
—Friendly Indulgences.—Open-Air Observing.—Method.—Perseverance.
—Definition in Towns.—Photography.—Publications.—Past and Future.
—Attractions of Telescopic Work.

Preparation.—An observer in commencing work in any department of astronomy will find it a very great assistance to his progress if he carefully reads and digests all that has been previously effected in the same line. He will see many of the chief difficulties and their remedies explained. He will further learn the best methods and be in the position of a man who has already gained considerable experience. If he enter upon a research of which he has acquired no foreknowledge he will be merely groping in the dark, and must encounter many obstacles which, though they may not effectually turn him from his purpose, will at least involve a considerable expenditure of time and labour. On the other hand, a person who relies upon guidance from prior experimentalists will probably make rapid headway. He will be fortified to meet contingencies and to avoid complications as they arise. He will be better enabled to discriminate as to the most eligible means and will confidently endeavour to push them to the furthest extent. By adopting existing instructions for his direction and familiarizing himself with the latest information from the best authorities he will in a great measure ensure his own success or at least bring it within measureable distance. The want of this foreknowledge has often been the main cause of failure, and it has sometimes led to misconceptions and imaginary discoveries ; for after much thought and labour a man will overcome an impediment or achieve an end in a way for which he claims credit, only to find that he has been anticipated years before and that

Fig. 18.

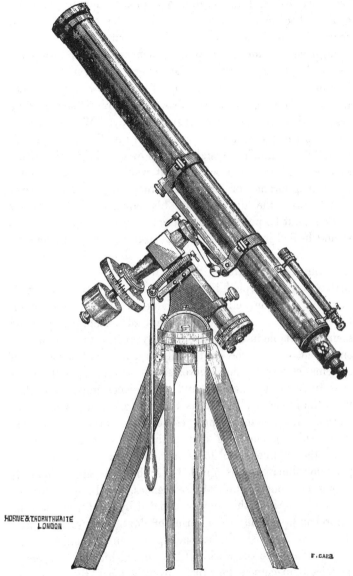

HORNE & THORNTHWAITE
LONDON

F. CARR

Refracting-Telescope on a German Equatoreal.

F 2

had he consulted past records, his difficulties would have been avoided and he might have pressed much nearer the goal.   Too much importance cannot be attached to the acquisition of foreknowledge of the character referred to, though we do not mean that former methods or results are to be implicitly trusted.   Let every observer judge for himself to a certain extent and let him follow original plans whenever he regards them as feasible ; let him test preceding results whenever he doubts their accuracy.   We recommend past experiences as a guide, not as an infallible precept.   It would be as much a mistake to follow the old groove with a sort of credulous infatuation as it would be to enter upon it in utter ignorance of theoretical knowledge.   An observer should take the direction of his labours from previous workers, but be prepared to diverge from acknowledged rules should he feel justified in doing so from his new experiences.

Working-Lists.—Full   advantage should be taken of good observing weather.   Sir John Herschel most aptly said that no time occupied in the preparation of working-lists is ill-spent.   In our climate the value of this maxim cannot be overrated.   If the 100 hours of exceptionally good seeing, available in the course of the year, are to be profitably employed, we must be continually prepared with a scheme of systematic work.   The observer should compile lists of objects it is intended to examine, and their places must be marked upon the globe or chart so as to avoid all troublesome references during the actual progress of observation.   If he has to consult ephemerides and otherwise withdraw attention from the telescope he loses valuable time : moreover the positions hurriedly assigned in such cases are frequently wrong and entail duplicate references, involving additional waste of time ; all this may be avoided by careful preparation beforehand.   If he has a series of double or variable stars to observe he must tabulate their places in convenient order so as to facilitate the work.   If he intend hunting up nebulæ or telescopic comets he must carefully mark their positions relatively to adjoining stars.   In the case of selenographical objects or planetary markings he may equally prepare himself by previous study.   Adopting these precautions, objects may be readily identified and the work expedited. When no such preparation is made much confusion and loss of time result. On a cloudy, wet day observers often consider it unnecessary to make such provision and they are taken at a great disadvantage when the sky suddenly clears. A good observer, like a good general, ought to provide, by the proper disposition of his means, against any emergency. In stormy weather valuable observations are often permissible if the observer is prompt, for the definition is occasionally suitable under such circumstances. The most tantalizing weather of all is that experienced during an anti-cyclone in winter. For a week or two the barometer is very steady at a high reading, the air is calm, and the sky is obscured with an impenetrable mass of clouds.

Wind.—The influence of wind on definition has been much discussed in its various aspects, but it is scarcely feasible to lay down definite rules on the subject. The east wind is rarely favourable to good seeing, but the law is far from absolute. We must remember that several distinct currents sometimes prevail, and the air strata at various elevations are of different degrees of humidity and therefore exercise different effects upon telescopic definition. A mere surface breeze from the east may underlie an extensive and moist current from the south-west, and telescopic definition may prove very fair under the combination. Calm nights when there is a little haze and fog, making the stars look somewhat dim, frequently afford wonderfully good seeing. As a rule, when the stars are sparkling and brilliant, the definition is bad ; planetary disks are unsteady and the details obliterated in glare. But this is not always so. I have sometimes found in windy weather after storms from the west quarter, when the air has become very transparent, that exceptionally sharp views may be obtained ; but unfortunately they are not without drawbacks, for the telescope vibrates violently with every gust of wind and the images cannot be held long enough for anything satisfactory to be seen. The tenuous patches of white cirrous cloud which float at high altitudes will often improve definition in a surprising manner, especially on the Moon and planets. Of course this does not apply to nebulæ or comets, which are objects of totally different character and essentially require a *dark* night rather than good definition before they may be seen under the best conditions. As a rule, a steady, humid atmosphere is highly conducive to good seeing, and it is rather improved than impaired by a little fog or thin, white cloud. Some unique effects of peculiar definition, such as oval or triangular star disks, have been occasionally recorded, but we must content ourselves with a bare reference to these phenomena. With regard to the general question it may, however, be added that the character of the seeing often varies at very short intervals in this climate. In the course of a night's work the definition will sometimes fluctuate in a most remarkable manner. An observer who comes to the telescope and finds it impossible to obtain satisfactory images should not entirely relinquish work at the first trial. After an interval he should again test its performance, for it frequently happens that a night ushered in by turbulent vapours, improves greatly at a later period, and in the morning part becomes so fine that it is worthy to be included in the select 100 hours assigned by Sir W. Herschel as the annual limit. Those who reside in towns will usually get the best definition after midnight, because there is less interference then from smoke and heated vapours: It would greatly conduce to our knowledge of atmospheric vagaries as affecting definition, if observers, especially those employing large aperture, preserved records as to the quality of the seeing, also direction of wind and readings of the barometer and thermometer.

Vision.—There are perhaps differences quite as considerable in powers of vision as in quality of definition. It is not meant by this that the same person is subject to great individual variations, though some people are certainly liable to fluctuations, according to state of health and other conditions. Some eyes, as already stated, are less effective in defining planetary markings than in detecting minute stars or faint satellites of distant planets. Of course the natural capacity is greatly enhanced by constant practice, for the human eye has proved itself competent to attain a surprising degree of excellence by habitual training. Frequent efforts, if not

overpressed so as to unduly strain the optic nerves, are found
to intensify rather than weaken the powers of sight. Thus a
distinguishing trait among astronomers has been their keen-
ness of vision, which, in many cases, they have retained to
an advanced age. It is true Dr. Kitchiner said his " eye at
the age of forty-seven became as much impaired by the
extreme exertion it had been put to in the prosecution of
telescope trials, as an eye which has been employed only in
ordinary occupations usually is at sixty years of age !—to
cultivate a little acquaintance with the particular and comparative powers of telescopes requires many extremely eye-
teasing experiments." But the Doctor's opinion is not generally confirmed by other testimony, the fact being that the
eye is usually strengthened by special service of this character.
To unduly tax or press its powers must result in injury ; but
it is well known that the capacities of our sight and other
senses are enhanced by their healthy exercise, and that comparative disuse is a great source of declining efficiency.
Before the observer may hope to excel as a telescopist it is
clear that a certain degree of training is requisite. Many
men exhibit very keen sight under ordinary circumstances,
but when they come to the telescope are hopelessly beaten
by a man who has a practised eye. On several occasions the
writer was much impressed with evidences of extraordinary
sight in certain individuals, but upon being tested at the
telescope they were found very deficient, both as regards
planetary detail and faint satellites. Objects which were
quite conspicuous to an experienced eye were totally invisible
to them. I believe it is a good plan for habitual observers
to employ method in exercising their sight. In my own case
I invariably use the right eye on the markings of planets
and the left on minute stars and satellites. Practice has
given each eye a superiority over the other in the special
work to which it has been devoted, and I fancy the practice
might be more generally followed with success.

It is an advantage to keep both eyes open when in the
act of observing, especially when surrounding objects are
perfectly dark and there is no distracting light from neighbouring windows or lamps. The slight effort required to

keep the disengaged eye closed interferes with the action of the other, and though this is but trivial, critical work is not efficiently performed under such conditions. Whenever light interferes the observer may exclude it by a shade so arranged as to afford complete protection to the unoccupied eye.

If faint objects are to be examined the observer should remain in a dark situation for some little time previously, so that the pupil of the eye may be dilated to the utmost extent and in a state most suitable for such work. After coming from a brilliantly lit apartment, or after viewing the Moon or a conspicuous planet, the eye is totally unfit to receive impressions from a difficult object, such as a minute star or faint nebula or comet ; some time must be allowed to elapse so that the eye may recover its sensitiveness. As a rule amateurs will find it best to confine their attention to one class of objects only on the same evening, for if the Moon is first examined and then immediately afterwards the telescope is directed upon double stars and nebulæ, the latter objects are little likely to be seen with good effect. If faint objects generally are persistently studied night after night and the observer refrains from solar and lunar work, his eye will acquire greater sensitiveness and he will readily pick up minute forms which are utterly beyond the reach of a man who indiscriminately employs his eye and telescope upon bright and faint objects.

Records.—With regard to records, every observer should make a note of what he sees, and at the earliest possible instant after the observation has been effected. If the duty is relegated to a subsequent occasion it is either not done at all or done very imperfectly. The most salient features of whatever is observed should be jotted down in systematic form, so as to permit of ready reference afterwards. It is useful to preserve these records in a paged book, with an index, so that the matter can be regularly posted up. The negligence of certain observers in this respect has resulted in the total loss of valuable observations. Even if the details appear to possess no significance, they should be faithfully registered in a convenient, legible form, because many facts deemed of no moment at the time may become of considerable importance. The observer should never refrain from such descriptions because he attributes little value to them. Some men keep voluminous diaries in which there is scarcely anything worth record; but this is going to the other extreme. All that is wanted is a concise and brief statement of facts. Some persons have omitted references to features or objects observed because they could not understand them, and rather distrusted the evidence of their eyes; but these are the very experiences which require careful record and reinvestigation.

Drawing.—Few observers are good draughtsmen; but it is astonishing how seldom we meet with real endeavours to excel in this respect. Every amateur should practise drawing, however indifferent his efforts may be. Delineations, even if roughly executed, are often more effective than whole pages of description. Pictorial representations form the leading attraction of astronomical literature, and are capable of rendering it more interesting to the popular mind than any other influence. They induce a more apt conception of what celestial objects are really like than any amount of verbal matter can possibly do. For this reason it becomes the obvious duty of every observer to cultivate sketching and drawing, at least in a rudimentary way. He will frequently find it essential to illustrate his descriptions, so as to ensure their ready comprehension. In fact, a thoroughly efficient observer must of necessity become a draughtsman. It should, however, be his invariable aim to depict just what he sees and in precisely the form in which it impresses his eye. Mere pictorial embellishments must be disregarded, and he should be careful not to include doubtful features, possibly existing in the imagination alone, unless he intends them simply for his own guidance in future investigations. If he sees but little, and it is faithfully delineated, it will be of more real value than a most elaborate drawing in which the eye and imagination have each played a part. It is an undoubted fact that some of the most striking illustrations in astronomical handbooks are disfigured by features either wrongly depicted or having no existence whatever. There is very great need for caution in representing such

markings only as are distinctly and unmistakably visible.
In all cases where the object is new or doubtful the observer
should await duplicate observations before announcing it.   It
is better that new features should evade discovery than that
delusive representations should be handed down to posterity.
As regards selenographical drawings I would refer the reader
to what Mr. Elger advises on p. 21 and 22 of volume v. of the
' Journal of the Liverpool Astronomical Society.'   My own
plan in sketching at the telescope is to first roughly delineate
the features bit by bit as I successively glimpse them, assuring
myself, as I proceed, as to general correctness in outline and
position ; then, on completion, I go indoors to a better light
and make copies while the details are still freshly impressed
on the mind.   To soften details a small piece of blotting-
paper must be wrapped round the pointed end of the pencil,
and the parts requiring to be smoothed gently touched or
rubbed until the desired effect is attained.   This simple
method, properly applied, will enable delicate markings to be
faithfully reproduced, and it certainly adds in no small degree
to the merit of a drawing.

Friendly Indulgences.—Every man whose astronomical pre-
dilections are known, and who has a telescope of any size, is
pestered with applications from friends and others who wish
to view some of the wonders of the heavens.   Of course it is
the duty of all of us to encourage a laudable interest in the
science, especially when evinced by neighbours or acquaint-
ances ; but the utility of an observer constituting himself a
showman, and sacrificing many valuable hours which might
be spent in useful observations, may be seriously questioned.
The weather is so bad in this country that we can ill spare an
hour from our scanty store.   Is it therefore desirable to
satisfy the idle curiosity of people who have no deep-seated
regard for astronomy, and will certainly never exhibit their
professed interest in a substantial manner ?   Assuredly not.
The time of our observers is altogether too valuable to be
employed in this fashion.   Yet it is an undisputed fact
that some self-denying amateurs are unwearying in their
efforts to accommodate their friends in the respect alluded to.
My own impression is that, except in special cases, the

observer will best consult the interests of astronomy, as well
as his own convenience and pleasure, by declining the character of showman ; for depend upon it a person who appre-
ciates the science in the right fashion will find ways and
means to procure a telescope and gratify his tastes to the
fullest capacity. Some years ago I took considerable trouble
on several evenings in showing a variety of objects to a
clerical friend, who expressed an intention to buy a telescope
and devote his leisure to the science. I spent many hours in
explanations &c.; but some weeks later my pupil informed
me his expenses were so heavy that he really could not afford
to purchase instruments. Yet I found soon after that he
afforded £30 in a useless embellishment of the front of his
residence, and it so disgusted me that I resolved to waste no
more precious time in a similar way.

Open-Air Observing.—Night air is generally thought to be
pernicious to health ; but the longevity of astronomers is
certainly opposed to this idea. Those observers who are
unusually susceptible to affections of the respiratory organs
must of course exercise extreme care, and will hardly be
wise in pursuing astronomical work out of doors on keen,
wintry nights. But others, less liable to climatic influences,
may conduct operations with impunity and safety during the
most severe weather. Precautions should always be taken to
maintain a convenient degree of warmth ; and, for the rest,
the observer's enthusiasm must sustain him. A "wadded
dressing-gown" has been mentioned as an effective protection from cold. I have found that a long, thick overcoat,
substantially lined with flannel, and under this a stout
cardigan jacket, will resist the inroads of cold for a long
time. On very trying nights a rug may also be thrown over
the shoulders and strapped round the body. During intense
frosts, however, the cold will penetrate (as I have found while
engaged in prolonged watches for shooting-stars) through
almost any covering. As soon as the observer becomes
uncomfortably chilly he should go indoors and thoroughly
warm his things before a fire. He may then return fortified
to his work and pursue it for another period before the frost
again makes its presence disagreeably felt. On windy nights

a knitted woollen helmet to cover the head, and reaching to the shoulders, is an excellent protection ; but an observer had better not wear it more often than is imperative, or it becomes a necessity on ordinary nights. It is a great mistake to suppose that "a glass of something hot" before going into the night air is a good preventive to catching cold. It acts rather in the contrary way. The reaction after the system has been unduly heated only renders the observer more sensitive, and the inhalation of cold air is then very liable to induce affections of the throat.

A telescope permanently erected in the open, and exposed to all weathers, must soon lose its smart and bright appearance, but it need lose none of its efficiency, which is of far more importance ; for it is intended for service, not for show. The instrument should be kept well painted and oiled. I find vaseline an excellent application for the screws and parts controlling the motions, as it is not congelative like common oils. The observer, before a night's work and before darkness sets in, will do well to examine his instrument and see that it is in the best condition to facilitate work. Whole tribes of insects take up their habitation in the base or framework, and even in the telescope itself if they can effect a lodgment; and I have sometimes had to sweep away a perfect labyrinth of spiders' webs from the interior of the main tube. On one occasion I could not see anything through the finder, try how I would. I afterwards discovered that a mason-wasp (*Odynerus murarius*) had adopted the vacuity in front of the eye-lens as a suitable site for her nest ; and here she had formed her cells, deposited her eggs, and enclosed the caterpillars necessary for the support of the young when hatched. On another night I came hurriedly to the telescope to observe Jupiter with my single-lens eyepiece, power 252, but could make nothing out of it but a confused glare, subject to sudden extinctions and other extraordinary vagaries. I supposed that the branches of a tree, waving in the wind, must be interposed in the line of sight, but soon saw this could not possibly be the explanation. Looking again into the eyepiece, I caught a momentary glimpse of what I interpreted for the legs of an insect magnified into

Fig. 19.

The Author's Telescope : a 10-inch With-Browning Reflector.

gigantic proportions and very distinct on the bright background formed by Jupiter much out of focus.   On detaching
the eyepiece and carrying it indoors to a light, an innocent-
looking sample of the common earwig crawled out of it.
The gyrations of the insect in its endeavours to find a place
of egress from its confinement had clearly caused the effects
alluded to.   Telescopic observers are thus liable to become
microscopic observers before they are conscious of the fact,
and perhaps also in opposition to their intention.   Other
experiences might be narrated, especially as regards nocturnal observing in country or suburban districts, where
the " serious student of the skies" may, like myself, find
diversion to his protracted vigils by the occasional capture
of a too-inquisitive hedgehog or some other marauding
quadruped.

Method.—Nearly all the most successful observers have
been men of method.   The work they took in hand has
been followed persistently and with certain definite ends
in view.   They recognized that there should be a purpose
in every observation.   Some amateurs take an incredible
amount of pains to look up an object for the simple satisfaction of seeing it.   But seeing an object is not observing
it.   The mere view counts for nothing from a scientific
standpoint, though it may doubtless afford some satisfaction
to the person obtaining it.   A practical astronomer, with
his own credit at stake and the interests of the science
at heart, will require something more.   In observing a
comet he will either fix its position by careful measurement with reference to stars near, or critically examine its
physical peculiarities, or perhaps both.   In securing these
data he will have accomplished useful work, which may
quite possibly have an enduring value.   In other branches
of observation his aim will be similar, namely to acquire
new materials with regard to place or to physical phenomena, according to the nature of the research upon which
he happens to be engaged.   Such results as he gathers are
neatly tabulated in a form convenient for after comparisons.
There have been instances, we know, where sheer carelessness
has resulted in the loss of important discoveries.   Lalande

must have found Neptune (and mathematical astronomy
would have been robbed of its greatest triumph) half a
century before it was identified in Galle's telescope, but
his want of care enabled it to elude him just when he was
hovering on the very verge of its discovery. Numerous
other instances might be mentioned. Failure may either
arise from imperfect or inaccurate records, from a want of
discrimination, from neglect in tracing an apparent dis-
cordance to its true source, or from hesitation. I may be
pardoned for mentioning a case within my own experience.
On July 11, 1881, just before daylight, I stood contem-
plating Auriga, and the idea occurred to me to sweep the
region with my comet eyepiece, but I hesitated, thinking
the prospect not sufficiently inviting. Three nights later
Schæberle at Ann Arbor, U.S.A., discovered a bright telescopic comet in Auriga! Before sunrise on October 4 of
the same year I had been observing Jupiter, and again
hesitated as to the utility of comet-seeking, but, remem-
bering the little episode in my past experience, I instantly
set to work, and at almost the first sweep alighted upon
a suspicious object which afterwards proved itself a comet
of short period. These facts teach one to value his opportunities. They cannot be lightly neglected, coming as they do
all too rarely. The observer should never hesitate. He must
endeavour to at least effect a little whenever an occasion offers ;
for it is just that little which may yield a marked success—
greater, perhaps, than months of arduous labour may achieve
at another time.

Perseverance.—Persistency in observation, apart from the
value derived from cumulative results, increases the powers
of an observer to a considerable degree. This is especially
the case when the same objects are subjected to repeated
scrutiny. A first view, though it may seem perfectly satisfactory in its conditions and results, does not represent what
the observer is capable of doing with renewed effort. Let
us suppose that a lunar object with complicated detail is to
be thoroughly surveyed. The observer delineates at the first
view everything that appears to be visible. But a subsequent effort reveals other features which eluded him before,

and many additional details are gradually reached during
later observations. Ultimately the observer finds that his
first drawing is scarcely more than a mere outline of the
formation as he sees it at his latest efforts. Details which
he regarded as difficult at first have become comparatively
conspicuous, and a number of delicate structures have been
exhibited which were quite beyond his reach at the outset.
The eye has become familiarized with the object, and its
powers fairly brought out by training and experience. This
training is very serviceable, but is seldom appreciated in the
degree of its influence. Many a tyro has abandoned a projected series of observations on finding that his initiatory
view falls wofully short of published drawings or descriptions. He considers himself hopelessly distanced, and regards
it as impossible to attain—much less excel—the results
achieved by his predecessors. He does not realize that their
work is the issue of years of close application, and that it
represents the collective outcome of many successive nights.
I need hardly say that it is a great mistake to anticipate
failure in this way. No telescopic work has been done in the
past that will not be done better in the future. No observer
can rate his capacity until he has rigorously tested it by
experience. The eye must become accustomed to an object
before it is able to do itself justice. Those who have been
sedulously engaged in a certain research will, as a rule, see
far more than others who are but just entering upon it—not
from a natural superiority of vision, but because of the aptitude
and power acquired by practice. No matter how meagre an
observer's primary attempts may be, he should by no means
relax his efforts, but rather feel that his want of success
must be remedied by experience. It is a common fault with
observers that they leave too much to their instruments, and
rely upon them for the results which really depend entirely
upon their personal endeavours. A skilled workman will do
good work with indifferent tools; for after all it is the
character of the man that is evident in his results, and not so
much the resources which art places in his hands.

Much also depends upon the feelings by which the amateur
is actuated when he commences work. A few enter into it

with a degree of energy and determination that knows no wearying and will accept no defeat. Others display a half-hearted enthusiasm, and are constantly doubting either their personal ability or their instrumental means. Many others, again, when the circumstances appear a little against them regard failure as inevitable. It need hardly be said, however, that every difficulty may be surmounted by perseverance, and that a man's enthusiasm is often the measure of his success, and success is rarely denied to him whose heart is in his work.

Definition in Towns.—The astronomical journals contain some interesting references to the definition of telescopes in large towns. Of course the purer the air the better for observational purposes. But observers who reside in populous districts need not despair of doing really useful work. The vapours hanging over a large city are by no means so objectional as is commonly supposed. When they are circulating rapidly across the observer's field of view they will prove very troublesome at times ; but in a comparatively tranquil state of the air definition is excellent. I have frequently found planetary markings very sharp and steady through the smoke and fog of Bristol. The interposing vapours have the effect of moderating the bright images and improving their quality. When there is a driving wind, and these heated vapours from the city are rolling rapidly past, objects at once appear in a state of ebullition, and the work of observation may as well be postponed. Smoke from neighbouring chimneys is utterly ruinous to definition : a bright star is transformed into a seething, cometary mass, and the planets undergo contortions of the most astonishing character. Large instruments being more susceptible to such influences—and, indeed, to atmospherical vagaries of all kinds—are chiefly affected by the drawbacks we have alluded to ; but there are many opportunities when their powers may be fully utilized. In sweeping for faint comets, or in other work (such as the observation of nebulæ) where a dark sky is the first essential, a town station has a manifest disadvantage because of the artificial illumination of the atmosphere. But for general telescopic work the conditions do not offer a serious impediment, espe-

cially if the observer is careful to seize the many suitable
occasions that must occur.   The direction of the wind relatively to his position and the central part of the city, will
occasion considerable differences to an observer who uses a
telescope in a suburban locality.

Photography.—Upon this branch of practical astronomy
not much will be said in this volume, as it is rather beyond
its scope, and possibly also beyond the resources of ordinary
amateurs, so far as really valuable work is concerned.   A
reference must, however, be made to an innovation which has
deservedly assumed a very prominent place, and is clearly
destined to exert an accelerating influence on the progress of
exact astronomy.   At present it is impossible to foretell how
far it may be employed and extended, but judging from recent
developments its applications will be as manifold as they will
be valuable.   Photographic records possess a great advantage
over others, because they are more accurate and therefore
more reliable.   They are pictures from Nature taken by
means free from the bias and error inseparable from mere
eye-estimations or hand-drawings.   The latter are full of
discordances when compared one with another, and can
seldom be implicitly trusted ; but in the photograph a
different state of things prevails.   Here we have a faithful
portrayal or reproduction of the object impressed by itself
upon the plate.   Hence it can be depended upon, because
there has been no intermediate meddling either with its
position or features by what may be termed artistic mis-
representation.   True, there may be imperfections in the
process ; trifling flaws and obstructions will invariably creep
in wherever comparatively new and novel work is attempted,
but these will but little detract from the value of its results.
Photography is obviously a means of discovery as well as a
means of accurate record ; for nebulæ and faint stars quite
invisible to the eye have been distinguished for the first time
upon the negatives.   Those of our amateurs who intend
working in this branch will find it a productive one, and
not decaying in interest ; but the necessary outfit will be
expensive if thoroughly capable instruments are to be
employed in the service.

STANMORE OBSERVATORY.
OUTSIDE VIEW

Publications.—The observer of to-day may esteem himself particularly fortunate in regard to the number and quality of the astronomical journals within his reach. Discoveries and current events receive prompt notice in these, and readers are fully informed upon the leading topics. Among the best of the periodicals alluded to are 'The Observatory' (Taylor & Francis, London), 'The Sidereal Messenger' (Northfield, Minn., U.S.A.), and *L'Astronomie* (Gautier-Villars, Paris). The *Astronomische Nachrichten* (Kiel, Germany) is a very old and valued serial, and 'The Astronomical Journal' (Cambridge, Mass., U.S.A.) may also be favourably mentioned. The 'Monthly Notices' of the Royal Astronomical Society and the 'Journals' of the Liverpool and British Astronomical Societies contain many interesting materials. 'Nature,' 'The English Mechanic,' and 'Knowledge' are among the English journals which devote part of their space to the science ; and the beautiful illustrations in the latter entitle it to special recognition. It is evident, from this short summary, the amateur will find that his literary appetite may be amply satisfied, and should he desire a channel for recording his own work or ideas the publications referred to offer him every facility and encouragement.

As to almanacks, the 'Nautical,' which has been termed "The Astronomer's Bible," includes a mass of tabular matter, some portion of which is of utility to the amateur, but it does not give data which are to be found in some other publications. I refer particularly to ephemerides of the satellites of Mars, Saturn, Uranus, and Neptune, to the dates of max. and min. of variable stars, to the times of rising and setting of the Sun, Moon, and planets, to the epochs and positions of meteor-showers, &c. The annual 'Companion to the Observatory' furnishes most of these details, and 'Whitaker's Almanack' and Brown & Sons' 'Nautical Almanack' each contain a large amount of serviceable information. The latter, however, is chiefly devoted to topics connected with Navigation, while 'Whitaker's Almanack' is an extensive repertory of general facts.

With respect to handbooks much depends upon the direction of the observer's labours, for he will obviously require

works dealing expressly with his special subject. As a
reliable companion to the telescope, Webb's ' Celestial Objects
for Common Telescopes ' (4th edit., 1881) is indispensable ;
as a work of reference, and one forming an exhaustive con-
spectus of astronomical facts, Chambers's ' Descriptive Astronomy ' (4th edit., in 3 vols., 1889) may be recommended.
Ledger's ' The Sun, its planets and their satellites ' is another
good descriptive work.   The beginner will find Noble's
' Hours with a 3-inch Telescope ' full of very instructive and
agreeable material ; while the more experienced astronomer,
requiring a masterly exposition of the principles of the
science, must procure Sir J. Herschel's ' Outlines ' (11th edit.,
1871).   In departmental work books of more exclusive character will be necessary.   Thus, students of solar physics will
want Young's volume on ' The Sun ;' observers of our satellite
will need Neison's ' Moon.'   Those who find double stars
interesting should get Crossley, Gledhill, & Wilson's ' Hand-
book ' and Chambers's revised edition of Admiral Smyth's
' Cycle ;'   others working on variable stars will need the
Catalogues of Chandler and Gore. Jovian phenomena are well
represented in Stanley Williams's ' Zenographic Fragments.'
Comets have been fully treated of in works by Cooper, Hind,
and Guillemin ; while to the observer of eclipses Johnson's
' Eclipses Past and Future ' is a valuable guide.   Everyone
interested in nebulæ will of course require Herschel-Dreyer's
' General Catalogue,' containing 7840 objects and published
by the Royal Astronomical Society in 1888.   As to planetary
observations, the several works of Webb, Chambers (vol. i.),
and Ledger, first cited, supply a large amount of detail,
almost obviating the necessity for further books.

Past and Future.—Observers and telescopes go on increasing
day by day, and the future of astronomy has a most brilliant
outlook.   Photography has latterly effected a partial revolution
in observation, though it can never entirely supersede old
methods.   Spectrum analysis, too, has formed a valuable
acquisition during the last quarter of a century.   With the
new and refined processes, and with the gigantic instruments
which have been erected, we may confidently anticipate many
additions to our knowledge, especially in regard to very small

and faint bodies which the inferior appliances of previous years have failed to grasp. And it is certain that some of the presumed discoveries of past times must be expunged, because not verified by the more perfect and powerful researches of a later date. Let us place in parallel columns (1) a few of the suspected objects thus to be erased, and (2) some of those which the future will probably add to our store :—

(1.)	(2.)
Satellite of Venus.	Satellites of Uranus and Neptune.
Vulcan.	Ultra-Neptunian Planet.
Active Volcanoes on the Moon.	Changes on the Moon.
Detached cusps of Venus and Mercury indicating high mountains.	Rotation of Mercury, Venus, Uranus, and Neptune.
Rings of Uranus and Neptune.	Minor Planets.
Multiple companions to Polaris and Vega.	Periodical Comets.
	Nebulæ and Double Stars.

Whatever may be the direction of future enquiries or the departures from old and tried methods, ordinary amateurs with small instruments, though handicapped more heavily as regards the prospect of effecting discoveries, may yet always be expected to accomplish useful work. Even to him who simply makes the science a hobby and a source of recreation in a leisure hour after the cares of business, the sky never ceases to afford a means of agreeable entertainment. He may neither achieve distinction nor seek it ; but this he will assuredly do—gain an instructive insight into the marvellous works of his Creator, and acquire a knowledge which can only exercise an elevating tone to his life. The observer who quietly, from his cottage window, surveys the evening star or the new Moon through his little telescope often finds a deeper pleasure than the proficient astronomer who, from his elevated and richly appointed observatory, discovers new orbs with one of the most powerful instruments ever made.

Attractions of Telescopic Work.—In concluding our comments we may briefly refer to the importance and pleasure attached to telescopic work, and the growing popularity of observation in the attractive and diverse field of astronomy. A telescope may either be employed as an instrument of scientific discovery and critical work, or it may be made a source of recreation and instruction. By its means the powers of the eye are so far assisted and expanded that we are enabled to form a clearer conception of the wonderful works of the Creator than could be obtained in any other way. Objects which appear to natural vision in dim and uncertain characters are resolved, even in telescopes of the smallest pretentions, into pictures of well-defined outlines containing details of configuration far exceeding what are expected. And it is entirely owing to the exact measurements obtained under telescopic power that many of the most important problems of astronomy have been satisfactorily solved. To this instrument we are indebted, not only in a great measure for our knowledge of the physical features of many celestial bodies, but also for the accurate information we have gained as to their motions, distances, and magnitudes. Apart from this it is capable of affording ample entertainment to all those who are desirous of viewing for themselves some of the absorbing wonders of astronomy as described in our handbooks. And a demonstration of this practical kind is more effective than any amount of description in bringing home to the comprehension of the uninitiated the unique and picturesque side of astronomy.

CHAPTER V.

THE SUN.

Solar Observations.—Early notices of Spots.—Difficulties of the old observers.—Small instruments useful.—Tinted glass.—Solar Diagonal.—Structure of a Spot.—Methods of Drawing.—Ascertaining Dimensions.—Observer's aims.--Eclipses of the Sun.—Periodicity of Spots.—Crateriform structure.—"Willow-Leaves."—Rotation of the Sun.—Planetary bodies in transit.—Proper motion of Sun-spots.—Rise and decay of Spots.—Black Nuclei in the umbræ.—Bright objects near the Sun.—Cyclonic action.—Sudden outbursts of Faculæ.—Shadows cast by Faculæ.—Veiled Spots.—Recurrent disturbances.—Recurrent forms.—Exceptional position of Spots.—The Solar prominences.

> " Along the skies the Sun obliquely rolls,
>   Forsakes, by turns, and visits both the poles;
>   Diff'rent his track, but constant his career,
>   Divides the times, and measures out the year."

THE Sun is not an object comprehended in the title of this volume. But to have omitted reference to a body of such vast importance, and one displaying so many interesting features to the telescopic observer, would have been inexcusable. We may regard the Sun as the dominant power, the controlling orb, and the great central luminary of our system. The phenomena visibly displayed on his surface assume a particular significance, as affecting a body occupying so high a place in the celestial mechanism.

The mean apparent diameter of the Sun is 32′ 3″·6, and his real diameter 866,000 miles. The apparent diameter varies from a minimum of 31′ 32″ at the end of June to a maximum of 32′ 36″ at the end of December ; and the mean value is reached both at the end of March and September. The Sun's mean distance from the Earth is about 92,900,000 miles, computed from a solar parallax of 8″·8, which appears to agree with the best of recent determinations. At this distance the linear value of 1″ of arc is 447 miles.

The Sun's apparent diameter is as follows on the first day
of each month :—

Jan. 1	32 36·0	July 1	31 32·0
Feb. 1	32 31·8	Aug. 1	31 35·8
Mar. 1	32 20·4	Sept. 1	31 47·0
April 1	32 3·8	Oct. 1	32 2·6
May 1	31 48·0	Nov. 1	32 19·2
June 1	31 36·4	Dec. 1	32 31·6

Solar observations may be pursued with a facility greater
than that attending work in some other departments of
practical astronomy. The Moon, planets, and stars have to
be observed at night, when cold air, darkness, and other circumstances are the cause of inconvenience ; but the student
of the Sun labours only in the light and warmth of genial
days, when all the incidentals to observation may be agreeably performed. There are, however, some drawbacks even
in this pleasant sphere of work. The light of the Sun is
so great that much persistent observation is apt to have
an injurious effect on the eye, and will certainly deaden its
sensitiveness on faint objects. In the summer months the
observer experiences discomfort during a lengthy observation
from remaining so long in the powerful rays of the Sun, some
of which must fall upon his face unless measures are adopted
to shield it. During the progress of solar work the student
should always provide for himself as much shelter as possible
from the glare, which must otherwise disturb that equanimity
of feeling in the absence of which no delicate research is
likely to be successfully conducted.

"Spots on the Sun" were remarked long before the telescope came into service. In the early Chinese annals many
references are made to these objects ; thus, in A.D. 188,
February 14, it is recorded—"The colour of the Sun reddish-
yellow ; a fleckle in the Sun (bird-shaped)." Other ancient
notices compare the spots to a flying bird, an apple, or an
egg. Many spots were seen in later years, especially in 321,
807, 840, 1096, &c. In 807 a large black spot upon the Sun
was watched during a period of eight days. It reflects much
credit upon observers of a past age that they performed so

many useful feats of observation, though relying simply upon
the powers with which Nature alone had endowed them.
They anticipated the telescope in some important discoveries.
Large sun-spots are not, it is true, difficult features to
perceive with the naked eye under certain circumstances ;
for whenever there is a fog or haze sufficiently dense to veil
the lustre of the Sun in suitable degree, they can be readily
seen, presuming, of course, that such spots are in existence at
the time. They are sometimes observed, in a purely casual
way, by people who may happen to glance at the Sun when
he is involved in fog and looks like a dull, red ball suspended
in the firmament. On one occasion, near sunset, in the
autumn of 1870, I saw four large spots on different parts of
the Sun, and these phenomena were very numerous at about
this time. When spots attain a diameter of $50''$ or more
they may be detected by persons of good sight; but if the
Sun is high and clear, coloured glass must be used to defend
the eye.

Doubt hangs over the question as to the first telescopic
observer of the spots. It is certain that Fabricius, Galilei,
Harriot, and Scheiner all remarked them in about the year
1611; and of these Fabricius perhaps deserves the chief
praise, as the first who published a memoir on the subject.
Galilei appears undoubtedly to have had priority in recognizing the bright spots, or *faculæ*. Scheiner discovered that
the black spots, or *maculæ*, are composed of a dark umbra and
a fainter outlying shade, called the penumbra. Arago quotes
him as having also described the Sun as "covered over its
whole surface with very small, bright, and obscure points, or
with lively and sombre streaks of very slender dimensions,
crossing each other in all directions." He announced, too,
that the spots were confined to a narrow zone on the north
and south sides of the equator, and this he termed the "Royal
Zone."

Some grave difficulties appear to have marked the attempts
of the earlier observers ; for they did not all use coloured
glasses, and the dazzling light of the Sun, intensified by their
lenses, often overpowered the sight, and so we find them
awaiting opportunities when fog partly obscured the Sun

near his rising or setting. Thus Harriot, who seems to have noticed and figured three sun-spots as early as 1610, Dec. 8, says :—" The altitude of the Sonne being 7 or 8 degrees, and it being a frost and a mist, I saw the Sonne in this manner." His drawing followed. On another occasion he says :—" A notable mist : I observed the Sonne at sundry times, when it was fit." Fabricius advised other observers to commence their observations by admitting only a small portion of the Sun into the field, so that the eye might be prepared to receive the light of the entire disk. Galilei was equally unaware of the advantage of tinted glass, and adopted the expedient of scanning the Sun when placed in the vicinity of the horizon. He remarks that " the spot of 1612, April 5 appeared at sunset ;" and his writings contain other references of similar import. Scheiner, however, appears to have been more alive to the requirements of the work, and employed a plain green glass placed in front of the object-lens of his telescope.

Under the various circumstances we have been alluding to, the views obtained of the solar surface must necessarily have been of a very defective character, and the old observers at least deserve our sympathy in their exertions. No such obstacles confront the observer now. He has everything provided for him. Instrumental devices rob the Sun of his noonday brilliancy, and the eye serenely scans the details of his expansive image without the slightest pain or effort.

Small telescopes are peculiarly well adapted for solar observations. A good 3-inch refractor or 4-inch reflector will reveal an astonishing diversity of structure in the spots, and show something of the complicated *minutiæ* of the general surface. If the aperture of either instrument is 2 inches more than that stated, so much the better ; but further than this it is rarely advisable to go. When the objective or mirror exceeds a diameter of 5 or 6 inches a stop often improves the images, and even smaller instruments will perform better when a little contracted. Definition is here the point to be desired ; of light we have a superabundance. But if the observer meditates a critical analysis of the detail, either of a single spot, of a group of spots, or of a small area of the luminous surface, then a fair amount of aperture should be used, because greater aperture means greater separating power, and the latter will be useful in resolving the network of fibrous materials of which apparently the whole surface is composed. But for the common requirements of the observer an instrument of 3 or 4 inches will be found very effective, and it can either be used on a short tripod stand, placed on a steady table near a window having a south aspect, or it may be mounted on a tall garden stand and, according to the owner's pleasure, either fixed at his window or in his garden. Two powers will be really necessary—one of about 60 and a field of quite 33″ to contain the entire disk and give a good general view, and another of 150 to which the observer will have recourse when examining details. Additional eyepieces will be sometimes useful, especially one of about 100 ; but the power of 60 previously recommended will, if a Huygenian, answer the same purpose, for if the field-lens is removed it will be increased to about 90. And should the observer think that anything is to be gained by a higher magnifier than 150, let him use the eye-lens only of that power. I have obtained many exquisite views of sun-spots with a single lens, and, instead of purchasing new eyepieces, a real advantage will be derived in adopting the plan suggested. There will be a smaller field and more colour about the image, but the improvement in definition is considerable, and more than balances these disadvantages.

Tinted glass must always be employed, unless a dense fog prevails, in which case the example of the old observers may be emulated. Several coloured glasses, of various depths, are needed for use according as the occasion requires. With a high Sun on a bright June day a darker tint will be necessary than in the winter, when the Sun's rays are but feebly transmitted through the horizontal vapours. Red glass is unsatisfactory, as there is much heat and glare with it ; but when used in combination with green the effect is excellent. Green alone is often used, and answers well ; but it is not always thick and dense enough for the purpose. The plan of Sir W. Herschel, to interpose a glass trough of diluted ink, has never become popular, though he found it to succeed admirably.

Smoked glass is also adapted for solar work, and recommends itself as being always obtainable at a minute's notice. Some observers use a Barlow lens, with a thin film of silver deposited on the surfaces. It is then sufficiently transparent to give a neutral tint when held before a light, and sharp definition is said to be obtained without additional protection. Mr. Thornthwaite has also employed a coloured Barlow lens with effect.

A solar diagonal is a very necessary appliance if the observer would ensure perfect safety ; for any refractor exceeding 2-inches aperture may, when turned on the Sun, focus enough heat to fracture the tinted sun-glass. The diagonal, by preserving a part only of the solar rays which are transmitted by the object-glass, enables observations to be made in security. This little instrument is comparatively cheap, and no telescope is complete without one. Dawes's solar eyepiece serves the same purpose in a different manner, but it is an expensive luxury. In the latter construction there is a perforated diaphragm fixed near the eyepiece and so arranged that the quantity of admitted light may be modified consistently with the observer's wishes.

In reflecting-telescopes with glass mirrors, effective views of the Sun are obtainable by employing unsilvered mirrors : for sufficient light is reflected by the glass surfaces to form good images of solar detail.

What, perhaps, interferes more than any other circumstance with successful observation of the Sun, is the fact that the rays, falling upon the telescope and objects near, induce a good deal of radiation, the direct tendency of which is to impair the definition and give a rippling effect to the disk. This is sometimes present in such force that the spots are subject to an incessant commotion, which serves to obliterate their more delicate features. A shady place is best, therefore, for such work ; and if the observer leaves his telescope for a short time, intending to resume observations, it should never be placed broadside to the Sun, or the tube will get hot, and heated currents must be generated in the interior, to the ruin of subsequent views.

A large sun-spot consists of an apparently black nucleus, a

brown umbra, divided possibly by veins of bright matter or by encroachments of the penumbra which surrounds it. The latter is of much lighter tone than the umbra, though often similar in its general form. The outer edges of the umbra are serrated or scalloped by rice-grain protuberances. The inner region of the penumbra is much brighter than the outer, and the latter often exhibits quite a dusky fringe, induced by lines of dark material intervening with the brighter particles. The filaments forming the penumbra— often grouped in a radial manner with reference to the centre of a spot—would appear to be more widely separated near the outer border of the penumbra, and sufficiently so to allow sections of the umbral layer of the Sun to be observed through the interstices. The lighter tint of the interior part of the penumbra is stated to be due to contrast ; but this is a mistake. The difference is too definite and distinct to permit such an explanation. Mr. Maunder says "that usually (not invariably) the penumbra darkens towards the umbra, and that the phenomenon as ordinarily described is merely an effect of contrast." My own observations, however, appear to show that there is an actual difference of detail in the outer and inner portions of the penumbra, which gives a darker tone to the former.

In drawing the forms of sun-spots the observer must be expeditious, because of the variations which are quickly and constantly affecting them. In concluding a sketch I find it essential to make several alterations in it, owing to the changes which have occurred in the spots during the interval of a quarter of an hour or so since it was commenced. The details must be filled in consecutively, each one being the result of a careful scrutiny. When finished, the whole sketch should be compared with the object itself and amended if found necessary. The observer should also mark upon the sheet the measured or estimated latitude and longitude of the spot, and make a finished drawing from the basis of his sketch as soon as possible afterwards. At Stonyhurst Observatory excellent delineations of solar phenomena are made ; and the late Father Perry, who lost his life in the cause of science, thus described the method :—" On every fine day the image of the

Sun is projected on a thin board attached to the telescope, and a drawing of the Sun is made, 10½ inches in diameter, showing the position and outline of the spots visible. It is the first duty of the assistant who makes the drawings to note the position of the spots, and sketch their outlines. He then proceeds to shade in the penumbra and to draw the finer details, comparing the drawing from time to time by placing it alongside the projected image of the spots. The position of the faculæ is then filled in with a red pencil, so that the eye can at once recognize their grouping with respect to sunspots, and the other details drawn with a black pencil." The same astronomer also stated that, " as a general rule, careful drawings of the projected image of the Sun give much more satisfactory pictures of the solar surface than the photographs taken even at our best observatories. It is quite true that occasionally an exquisite photograph on an enlarged scale may be obtained, which exhibits features such as no pencil could portray as accurately, but rarely indeed will the photograph furnish all the details that a practised eye and hand, kept patiently at the sketch-board, will detect and faithfully describe. And the reason is not far to seek ; for any experienced observer knows that, even on the finest day, the definition is continually changing with the sky, and that it is only at comparatively rare moments we can expect those perfect conditions that enable the finest details to stand out sharply, as Schiaparelli expresses it, like the faintest lines of a steel engraving. A photograph may be accidentally taken during one of these exceptionally favoured moments ; but a patient draughtsman is almost sure to secure several of these best opportunities at each prolonged visit to his sketch-board. What would, therefore, be a great acquisition at present is a series of careful solar drawings, taken at short intervals of time, on days when characteristic spots are visible upon the Sun ; and this would be the surest way of adding much valuable information to that already possessed concerning the changes that take place in the solar photosphere."

With regard to ascertaining the dimensions of sun-spots, very precise results require accurate means of measurement and some mathematical knowledge. For the general purposes of the amateur, who will only want round numbers, simple methods may be adopted with success. I have used, on a 4-inch refractor, a graduated piece of plane glass, mounted suitably for insertion in the focus of the eyepiece, and marked with divisions $\frac{1}{200}$ of an inch apart. With power 65 I find the Sun's disk at max. distance covers 83 divisions of the graduated lens ; so that one division $= 22''\cdot8$, the Sun's min. diameter being $1892''$. Each division, therefore, is equal to 10,434 miles, the Sun's real diameter being 866,000 miles.

I viewed a large spot on June 19, 1889, and found its

Fig. 20.

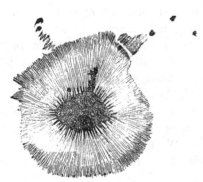

Sun-spot of June 19, 1889, 2ʰ P.M.

major axis covered $2\cdot6$ divisions, $= 59''\cdot3$ * ; so that its apparent length was about 27,000 miles. For

$$1892'' : 866,000 \text{ miles} :: 59''\cdot3 : 27,143 \text{ miles}.$$

The same method may be adopted if the image is thrown upon a screen.

Approximate values are to be obtained by means of fine

* The Rev. F. Howlett measured this spot on the following day, June 20, and found it $63''$ in its largest diameter. He used a small refractor, and projected the Sun's image on to a screen sufficiently distant for it to have a diameter of 3 feet.

cross wires fixed in the eyepiece. Note the exact interval occupied by the Sun in crossing the vertical wire, and also the interval occupied by the large spot or group. If the Sun is 133 seconds in passing the wire, and the group 6·5 seconds, then

133 seconds : 866,000 miles :: 6·5 seconds : 42,323 miles.

This plan is likely to be most successful when the Sun is near its meridian passage ; but it may be applied at any hour, if care is taken to adjust the eyepiece so that the Sun's motion is precisely at right angles to the vertical wire. One other plan may be mentioned. Draw on cardboard, with compasses, a circle about 10 or 12 inches diameter, and divide this with 31 parallel lines. Subdivide each of the spaces into 5, less prominently marked. Then, during observation, keep both eyes open, and hold or fix the circular disk at a distance enabling it to coincide with the telescopic image of the Sun. By carefully noting how many divisions the group covers on the cardboard, its dimensions may be readily found, because one division will be equal to about 5410 miles. Of course these methods* are simply approximate, and only strictly applicable to objects not far removed from the central regions of the Sun, because the spots are portions of a sphere, and not angles subtended by a flat surface. When close to the E. or W. limbs, foreshortening is considerable, though the polar diameter of a spot is not affected by it then.

Presuming an observer to have his 3- or 4-inch telescope duly fitted with a solar diagonal and tinted glass, he may naturally ask, after his curiosity has been satisfied by the contemplation of his first sun-spot, what he can do further :

---

* On May 13, 1890, at 3ʰ, I tested the three methods alluded to on a scattered train of small spots, and derived the following measurements of length :—

By glass micrometer	76,570 miles.
,, cross wires	76,610 ,,
,, cardboard disk	75,770 ,,

In this comparison I used an excellent 4-inch Cooke refractor, belonging to a friend.

What special features is he to look for? What changes
ought to be recorded? What are the doubtful points that
require to be cleared up as regards the Sun's physical
appearance? In what way are new and novel facts likely to
be glimpsed? In a word, he desires to know in what manner
he may employ his eyes and instrument usefully for science,
while also gaining pleasure for himself. Information like
this is often needed by the young student, and sometimes
indeed by men who have already gained a little experience, and
who possess much larger instruments than we have intimated
above. In endeavouring to offer suggestions in response to
such inquiries, I would remark that the nature and direction
of a research essentially depend upon several conditions, viz.
the observer's inclination, his instrumental equipment, his
place of observation, and the amount of time he can devote to
the pursuit of his object. There are very few men who, like
Schwabe of Dessau, will confront the Sun on nearly every day
for more than forty years in order to learn something of its
secrets. Such extraordinary pertinacity is fortunately not
required, except in special cases. Amateurs may effect much
valuable work in the short intervals which many of them steal
either from business or domestic ties and offer at the shrine
of astronomy.

There are quite a considerable number of attractive phenomena and features on which the solar observer will find
ample employment, and to the principal of these it may be as
well to make individual references.

Eclipses of the Sun.—These phenomena deservedly rank
amongst the most important and impressive events displayed
by the heavenly bodies, and they are specially interesting to
the possessors of small telescopes. Solar eclipses have been
so often made the subject of observation and discussion, that
our knowledge of the appearances presented may be considered to be nearly complete. The various aspects of Nature
on such occasions have been so attentively studied in their
manifold bearings, that virtually nothing remains for the
ordinary observer but to reexamine and corroborate facts
already well ascertained. He can expect to glean few
materials in a field where a plentiful harvest has just been

Fig. 21.

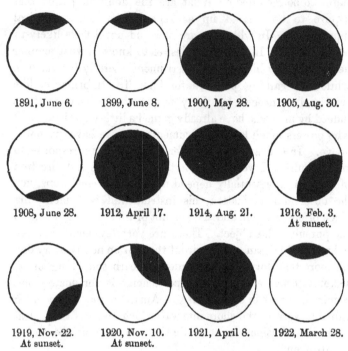

1891, June 6.   1899, June 8.   1900, May 28.   1905, Aug. 30.

1908, June 28.   1912, April 17.   1914, Aug. 21.   1916, Feb. 3.
                                                    At sunset.

1919, Nov. 22.   1920, Nov. 10.   1921, April 8.   1922, March 28.
At sunset.   At sunset.

Solar Eclipses visible in England, 1891 to 1922.

Fig. 22.

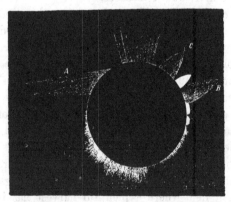

Total Solar Eclipse of August 19, 1887.

reaped. But the eclipsed Sun, if it has revealed most of its secrets to previous investigators, has certainly not declined in attractiveness ; and the amateur will find the spectacle still capable of exhibiting features which, though not full of the charms of novelty, will be sufficiently striking and diversified to be remembered long after the event has passed.

Eclipses recur in cycles of 18 years and 10 days (= 6585 days). This period was determined by the ancients, and called the *saros.* By its means the times and magnitudes of eclipses were roughly computed long before astronomy became an exact science.

A solar eclipse is really an *occultation* of the Sun by the Moon ; for the word *eclipse,* in its usual reference, denotes the obscuration of one body by its immersion in the shadow of another. During any single year there are never less than two eclipses, nor more than seven. Whenever there are two only, both are solar.

Since the fine solar eclipse of December 22, 1870, no large eclipse of the Sun has been visible in England. It is remarkable that during the thirty years from 1870 to 1900 these phenomena are all of an unimportant, minor character. Within the thirty years following 1891 there will be twelve solar eclipses, for which the Rev. S. J. Johnson has given projections (as shown on p. 98) for the period of greatest obscuration.

Total eclipses are extremely rare as regards their visibility at a given station. Thus between 878 and 1715 not one was observed at London, and during the next 500 years there will be a similar absence of such a phenomenon. The observer of total eclipses must perforce journey to those particular tracts of the earth's surface over which the band of totality passes. On such occasions photography plays an important part ; and the corona, the red flames, the shadow-bands, and numerous other features become the subjects of necessarily hurried observation and record, for totality endures for very few minutes*.

* The maximum duration of totality, under every favouring circumstance, appears to be about 8 minutes. The great eclipse which occurred on August 18, 1868, maintained the total phase for nearly

As regards ordinary partial eclipses, amateurs usually find ample entertainment in noting the serrated aspect of the Moon's contour projected on the bright Sun. It is also interesting to watch the disappearance and reappearance of the solar spots visible at the time. Rather a low magnifying power, with sufficiently expansive field to include the entire disk, is commonly best for the purpose of these observations.

Periodicity of Spots.—This detail may be said to have been fully investigated. Schwabe and Wolf have accomplished much in this direction. A work of this kind must, by the nature of it, extend over many years and entail many thousands of observations. It is therefore more suited to the professional astronomer than to the amateur, whose attention is more or less irregular owing to other calls. The sun-spot cycle is one of about 11 years, during which there are alternately few and many spots on the Sun. There appear to be some curious fluctuations, disturbing the regular increase and decrease in the number of spots; and these variations are worthy of more attention. The following are the years of observed maxima and minima of sun-spot frequency:—

Maxima.	Minima.
1828.	1833.
1837.	1843.
1848.	1854.
1860.	1867.
1870.	1878.
1883–4.	1890 (?).

These phenomena have been rare during the past few years. The next maximum may be expected in about 1894, when

---

6 minutes 50 seconds in the Gulf of Siam. In reference to this eclipse, Dr. Weiss says:—"In the records of ancient eclipses there are to be found only two which may be compared in size with that of August 18, 1868, but none in which the totality lasted so long. The first of these is the eclipse of Thales (28 May, 585 B.C.), which is said to have been the first predicted, and to have terminated a bloody war between the Lydians and the Medes. The second was visible on June 17, 1433, in Scotland, and the time of its occurrence was long remembered by the people of that country as 'the black hour.'"

solar observers will probably have an abundance of new materials to study.

Crateriform Structure.—In 1769 Prof. Wilson, of Glasgow, while watching a sun-spot with a Gregorian reflecting-telescope, remarked that, as it approached near the limb, the penumbra became much foreshortened on the interior side. He inferred from this that the spots were cavities, and the idea has been generally accepted ; so that these objects are sometimes termed solar craters, and commonly regarded as openings in the luminous atmosphere of the Sun. But the conclusion appears to be based on data not uniformly supporting it. In 1886 the Rev. F. Howlett published some observations which " entirely militate against the commonly received opinion that the spots are to any extent sunk in the solar surface as to produce always those effects of perspective foreshortening of the inner side of the penumbra (when near the limb) which have been described in various works on astronomy." In a number of instances the penumbra is wider on the side nearest the Sun's centre, whereas the converse ought to be the case on the cavity theory. The fine sun-spot of July 1889 offered an example of this ; for when it was near the W. limb the W. side of the penumbra was obviously much narrower than the E. side, so that the appearance would indicate the object as an elevation rather than a depression. The observer should keep a register of the aspect of all pretty large spots near the limb, and note the relative widths of the E. and W. sides of the penumbra. An extensive table of such results would be interesting, and certain to throw some light on the theory of spot-structure. It is of course possible that occasionally the inner side of the penumbra is broader than the outer, and thus appears wider even on the limb, though really forming the side of a shallow depression.

" *Willow-Leaves.*"—In 1861 the late Mr. Nasmyth announced that the entire solar surface was composed of minute luminous filaments in the shape of " willow-leaves," which interlaced one another in every possible variety of direction. This alleged discovery only met with doubtful corroboration. The objects were stated by some authorities to be simply identical with the " corrugations " and " bright nodules " of

Sir W. Herschel. Mr. Stone called them "rice-grains." The eagle-eyed Dawes thought "granulations" a more appropriate term, as it implied no consistency of form and size. Secchi referred to them as oblong filaments, and "rather like bits of cotton-wool of elongated form." The Rev. F. Howlett described the Sun as presenting a granulated, mottled appearance in a 3-inch Dollond refractor, and mentioned that on the morning of June 9, 1865, the aspect of its surface was like that of new-fallen snow, the objects "being not rounded but sharply angular." The opinions of observers were thus singularly diverse, and the result of several animated discussions at the Royal Astronomical Society was that little unanimity was arrived at, except as to the fact that the Sun's surface was crowded with small luminous filaments of elongated form, and either rounded or angular at the ends. There was no accord as to their precise forms or distinctive manner of grouping. Some of the observers averred that the "willow-leaves" or "rice-grains" had no title whatever to be regarded as a new discovery, the same appearances having been recognized long before. Gradually the contention ceased, and though more than a quarter of a century has passed since the discussion arose there has been little new light thrown on the subject.

Amateurs will therefore do well to probe deeper into this promising branch of solar observation. As Mr. Nasmyth himself stated, considerable telescopic power is required, combined with a good atmosphere. But comparatively small instruments will also be useful, because of their excellent definition and efficacy in displaying details on a brilliant orb like the Sun. A power of 150 should be employed in examining small regions of the general surface, and also the edges of the umbra and penumbra of sun-spots. When definition is unusually sharp, and the details very distinct, the magnifying power should be increased if it can be done with advantage; and the observer should utilize an occasion like this to the utmost extent. On a really excellent day more may be sometimes detected than during several weeks when the atmosphere is only moderately favourable. The observations, being of a critical nature, should not be

attempted in winter, when the Sun is low. I have frequently secured fine views of the delicate structure of the
solar surface between about 8 and 9 A.M. in the summer
months ; and this is often a convenient time for amateurs
to snatch a glimpse, before going to business.

With reference to the general question as to the existence
of the " willow-leaves," my conception of the matter is that
the features described by Mr. Nasmyth are not new. His
drawing of a spot in Sir J. Herschel's ' Outlines ' and
Chambers's ' Descriptive Astronomy ' exhibits objects extremely uniform in shape and size, and this uniformity I
have never observed in the penumbra of spots. As to the
engraving in the ' Outlines,' showing the aspect of the interlaced " willow-leaves " on the general surface, this is also
not realized in observation. The " corrugations " and
" bright nodules " of Sir W. Herschel aptly represent what
is seen, and they are possibly identical with the " very small
bright and obscure points " and " lively and sombre streaks "
of Scheiner, though seen much better and in more profusion
of detail through the improved modern telescopes. The so-
called " willow-leaves " are rounded at the ends, and are
consistent neither in size nor shape. They encroach upon
the umbra of the spots, and give a thatched appearance to
the edges. The penumbra also shows this in its outer limits,
where it is also fringed with lenticular particles. Drawings
by Capocci and Pastorff seventy-five years ago, and published in Arago's ' Popular Astronomy,' show the thatching
at the edges of the umbra quite as palpably as it is represented in recent drawings.

Rotation of the Sun.—By noting when the same individual
spots return to the same relative places on the disk, the
approximate time of rotation is easily deduced. This varies
according to the latitude of the spots* ; whence it is evident
the solar atmosphere is affected by currents of different velocities, causing the spots to vary in their longitudes with

* Carrington found that spots near the equator gave a shorter
rotation-period than those far removed from it. This offers an analogy to the spots on Jupiter, which move with greater celerity near
the equator, though the rule is not absolute.

reference to each other. The Earth's motion round the Sun causes the spots to travel apparently more slowly than they really do ; for observations prove that a spot completes a rotation in 27 days 5 hours, whereas the actual time, after making allowance for the earth's orbital motion, is about 25 days $7\frac{3}{4}$ hours. The period of rotation may be roughly found as follows, supposing a spot to return to precisely the same part of the disk in 27 days 5 hours :—

$$365^{d}\ 5^{h}\ 49^{m} + 27^{d}\ 5^{h} = 392\ \ 10^{h}\ 49^{m}.$$

Then

$$392^{d}\ 10^{h}\ 49^{m}\ (=565{,}129^{m}) : 365^{d}\ 5^{h}\ 49^{m}\ (=525{,}949^{m})$$
$$:: 27^{d}\ 5^{h}\ (=39{,}180^{m}) : 25^{d}\ 7^{h}\ 44^{m}\ (=36{,}464^{m}).$$

For exact results several circumstances have to be considered, such as the direction of the spot-motions across the disk, as the chords vary according to the season ; thus in June and December the spots traverse straight lines, while in March

Fig. 23.

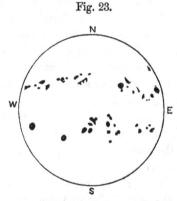

Belts of Sun-spots, visible October 29, 1868.

and September their paths are curved, like a belt on Saturn when the planet is inclined. Some of the spots display considerable proper motion ; so that it is best to observe a number of these objects, and reduce the times to a mean result. They are not very durable, rarely lasting longer than a few weeks ; but some of the more extensive disturbances are sustained for several months, during which many singular changes are effected. The period of rotation, as determined by several observers, is as follows :—

		d	h	m
1678.	Cassini	25	13	55
1718.	Bianchini	25	7	48
1775.	Delambre	25	0	17
1841.	Laugier	25	8	10
1846.	Kysæus	25	2	10
1852.	Böhm	25	12	29
1863.	Carrington	25	9	7
1865.	Schwabe	25	5	0
1868.	Spörer	25	5	31
1888.	Wilsing	25	5	47

The motion of rotation is similar in direction to that in which the planets move round the Sun, namely from west to east. Hence the spots come into view on the east limb of the Sun, and disappear at the west.

Planetary Bodies in transit.—During observation the observer should particularly watch any very dark, small spots that may be visible, such as are isolated and pretty circular and definite in outline. If an object of this character is seen it should be examined with a high power, and its aspect critically noted. Should the observer entertain any suspicion of its being of a planetary nature, he should carefully determine its position on the disk, and, after a short interval, reobserve it for traces of motion. If it remains stationary, its true solar origin will be proved. If motion is shown, then the successive positions of the object during its transit, and its place of egress, with the time of each observation, should be recorded. In such a case it would be a good plan to project the Sun's image, and mark the place of the suspicious object and chief sun-spots at short intervals. This would be more accurate than mere eye-estimation. The observer who scans the solar surface for intra-Mercurial planets must remember that, if any such bodies exist, they will probably be very diminutive. Venus, when on the Sun in December 1882, was a spot 63″ in diameter, and easily perceptible to the naked eye. Mercury, at the transits of 1861, 1868, and 1881, was a little less than 10″, but in 1878 was 12″. If "Vulcan," the suspected interior planet, has any existence it may possibly be much smaller than Mercury, and will thus escape observation, unless the observer exercises great care in the search. The mobile, planetary spots asserted to have been seen on the Sun in past years prove nothing definite, and appear to have been illusory.

Proper Motion of Sun-spots.—This feature is one deserving more investigation. The distances separating individual spots should either be measured with a micrometer or determined by transits across a wire, and the displacement recorded from hour to hour or from day to day. Spots in different latitudes will almost certainly exhibit some change of relative place ; and objects in the same latitude must be watched, for similar variations probably affect them. The physical peculiarities of such spots should be remarked, and also the alterations of appearance they undergo during the time they approach or recede from each other.

Rise and Decay of Spots.—Occasionally large spots are formed in an incredibly short time, and the disappearance of others has been equally sudden. Schwabe found, from many observations, that the western spots of a group are obliterated first ; but authorities differ. I have usually observed that the smaller, outlying members of a group vanish before the larger spot, which then contracts and is invaded by tongues of faculæ ; so that its effacement soon follows, and nothing remains to indicate the disturbance but bright ridges of faculæ, which are very conspicuous near the limb.

Black Nuclei in the Umbræ.—Dawes was the first to announce that the umbra sometimes included a much darker area or nucleus. This is present in nearly all large spots. A part of the umbra seems covered or veiled by a slightly luminous medium, and the portion unaffected looks black by contrast. On October 1, 1881, with a 2½-inch refractor, I saw a large sun-spot, the umbra of which was broken up into 7 fragments, and the S. preceding part appeared very black while the others showed a much lighter tint. In the fine spot of June 1889 a nucleus was also distinctly apparent ; and this feature is sometimes so obvious in large spots that it may be observed with an instrument of only 2-inches aperture. I have usually remarked the nucleus on one side of the umbra, and abutting the penumbra. It may be formed by light patches of transparent material floating over the umbra, and leaving a part free where the Sun's dark body is fully exposed. This light material is possibly suspended far above the umbra and inconstant in its position ; so that the place and form of the nucleus should always be noted for traces of change. It is necessary that such details should be closely watched during an entire day, or several days ; for the variations could then be followed, and perhaps reduced to some law. This persistence is very necessary, in order to solve many of the peculiarities of sun-spots, which, though pretty well known in appearance, have not been thoroughly studied in their various developments.

Bright Objects near the Sun.—Small, rapidly moving bodies have been occasionally reported as seen passing over the Sun. In several cases these have been prematurely assigned a meteoric origin. They have been described as luminous bodies of irregular shape, as moving in a common direction, and as being very distinct when projected on the dark sky just outlying the bright limb of the Sun. There is little doubt they are either the pappus of different kinds of seed, or convolutions of gossamer, which have been lifted to great heights in the air, and are rendered bright by reflection from the bordering Sun. In this connection I may mention some observations of my own with a 4-inch refractor :—

" 1889, MAY 20, 0^h 30^m P.M.—Bright points and little misty forms kept passing from the Sun's limb, at the average rate of 13 in a minute. They moved in the same direction as the clouds and wind. Some of them were followed by tails, which were far from straight. I saw them best when I focused the telescope for an object much nearer than the Sun. One of these forms would occasionally halt and pursue an irregular flight. It was evident they were terrestrial objects, with motions controlled by the wind.

" 3^h P.M.—Many bright objects still passing from the Sun's limb."

" 1889, MAY 22, 9^h A.M.—Observed vast numbers of

luminous particles floating about contiguous to the Sun's margin. They were clearly carried along by the wind; but this being very slight, their motions were extremely slow, and now and then many of them became nearly stationary. Their directions were far from uniform, though the general tendency was obviously in a common line of flight. I watched them for some time passing in a plentiful shower."

These objects are always noticed in summer-time, and I believe they would much more frequently attract remark but for the fact that they require a longer focus than the Sun and cannot be recognized when on the disk, to which the observer is usually giving the whole of his attention. Those who are often employed in solar work will find it an interesting diversion to look for these bodies. The instrument should be focused as for a distant terrestrial object, and only a part of the Sun's limb should be retained in the field of view of an eyepiece of moderately low power. Then, looking intently at the dark sky near the limb, the bright objects will be sometimes seen sailing past in considerable numbers.

Cyclonic Action.—The appearance in detail of certain spots, coupled with evidences of rotatory motion round their own centres, has induced the belief that they are liable to action in some degree similar to the cyclonic storms* which disturb and rend the terrestrial atmosphere. Such indications should be looked for in fairly conspicuous spots, and any peculiarities of the nature alluded to made the subject of close investigation. A spot showing features having a spiral tendency may not, however, have a gyratory movement about its centre. This can only be determined by critically noting the details, and frequently reobserving them for traces of motion. The penumbra always shows radiations converging on the umbra as a centre; but this is merely a form of structure, and proves nothing in evidence of a revolving storm.

Sudden Outbursts of Faculæ.—In September 1859 Carrington and Hodgson independently observed a striking out-

* In 1852 Dawes observed and measured a rotatory motion affecting a spot at the rate of about 17° per day.

burst of faculæ in front of a large group of spots which they were examining. It remained visible about five minutes, during which interval several patches of light travelled over a region nearly 34,000 miles in extent. An extraordinary magnetic disturbance was simultaneously recorded at the Kew Observatory, and sixteen hours afterwards there followed a magnetic storm of unusual severity. On another occasion Dr. Peters observed flashes of light cross and recross the umbra of a prominent spot with electric velocity. Some other startling observations of solar phenomena have been effected, and there is no question as to their having been matters of fact. In the presence of effects so sudden, so obvious, and so unexpected, no wonder the observers at first doubted the evidence of their eyes and suspected the cause to lie in a fractured glass or a fault of adjustment. But the corroboration afforded the clearest proof as to the actuality of the events described. They will doubtless occur again ; but these phenomena cannot be definitely predicted as to time, so that students of the solar surface should be prepared for a repetition of them whenever they may occur.

Shadows cast by Faculæ.—M. Trouvelot, while examining a large sun-spot on May 26, 1878, noticed that it was " completely surrounded by very brilliant and massive faculæ." " On one part of the penumbra an extraordinary appearance was perceived, which resembled so closely a shadow, as it would have been cast by the overhanging faculous mass, that it seemed useless to seek, and it was impossible to admit,

Fig. 24.

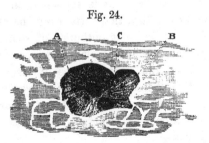

any other explanation. This shadow, the outline of which was a little diffused, had the same shape as, and reproduced

with great exactness, the outline of the faculous mass situated
above it. It was not so black as the opening in spots called
the umbra, but of a very dark tint." A similar feature was
seen by Kirk and Maclean on May 2, 1884, and the 'Observatory,' vol. vii. pp. 146, 170, and 197, contains some interesting particulars on this subject. Fig. 24 is a drawing by
Kirk, in which the shadow is represented by A, B ; at C " it
accurately followed the outline of the intensely white margin
of the spot."

Veiled Spots.—The late Father Perry described these
objects at the R.A.S. meeting on May 9, 1884, and said they
are to be seen all over the face of the Sun. They only exist
for two or three minutes, and then disappear. In one
instance he observed a train of these veiled spots stretching
over " a tenth part of the Sun's diameter, which was nearly
as obvious to the eye as the penumbra of an ordinary spot ;
it split into two throughout its whole length, and disappeared
in a minute. The veiled spots seem to be of two classes :
the one appear like small greyish clouds, which disappear after
a few minutes, as if they were formed and rapidly evaporated
by the Sun's heat, and the others seem to be connected with
the umbra of ordinary spots ; they appear about them, and
are more permanent than the ordinary veiled spots, lasting
sometimes two or three days, but never longer." These
markings appear to have been first detected by Trouvelot in
1875, and he gives some information as to this class of phenomena in the 'Observatory,' vol. viii. pp. 228 *et seq.*

Recurrent Disturbances *.—It is supposed, and with good
evidence affirming the idea, that certain regions of the Sun's
surface are subject to frequent outbursts of spots, which are
possibly due to forces acting from below the Sun's bright
atmosphere. After the disappearance of large groups or
isolated spots it is therefore advisable to watch the same
region for some time afterwards, to find whether it remains
perfectly quiescent, or whether it soon again becomes a seat
of activity and change.

* Lalande, in 1778, asserted that " there are spots of very considerable
magnitude, which reappear in the same physical points of the solar
disk. '

Recurrent Forms.—Certain spots observed at different times have exhibited appearances so nearly resembling each other that it has been considered the likeness may be due to something more than mere accident. Whenever such suggestive coincidences are recognized the observer should note them particularly, and secure drawings. It should be his aim to determine the exact intervals elapsing between the presentation of spots or groups of this character, and also whether they occupy the same latitude and longitude on the Sun's disk.

Exceptional Position of Spots.—The ordinary spots are rarely seen more than 35° distant from the solar equator or within 8° of it. They usually appear in the zones from 8° to 20° N. and S. of the equator. A few exceptions may be mentioned*. Mechain saw a spot in July 1780 having a latitude of 40½°; in April 1826 Capocci recorded one having 49° of S. latitude; Schwabe and Peters observed spots 50° from the equator. Lahire, in the last century, described a spot as visible in a latitude of 70°; but the accuracy of this observation has been questioned. Whenever a spot is seen near the equator, or very far removed from it, measures should be taken of its exact place; for it is desirable to learn something more of those disturbances which occasionally affect the more barren regions of the solar envelope.

The Solar Prominences.—Those amateurs who have included a spectroscope in their instrumental outfit will find the study of the chromosphere and prominences a most productive one. Huggins and Zöllner were the first to apply the "open-slit" method; and the study of the shape of the hydrogen prominences commenced in 1869. Tupman details ('Monthly Notices R.A.S.,' vol. xxxiii. p. 106) a series of observations which he secured in 1872 with a refractor of 3-inches aperture and a direct-vision spectroscope of five prisms. He mentions the cost of the entire apparatus as only £18, and says he entertains "no doubt that an equally effective instrument could be made for much less." The prominences

---

* A spot was visible on June 30, 1889, in 40° South latitude. Its recorded duration was 2 days. This object was observed at the Stonyhurst Observatory and at a station in North America.

appear to be of different kinds, and are known as "cloud-" and "flame"-prominences. Both are liable to rapid changes. Trouvelot, in June 1874, noticed "a gigantic comma-shaped prominence, 82,000 miles high, which vanished from before his eyes by a withdrawal of light as sudden as the passage of a flash of lightning." Since the study of these remarkable forms was rendered feasible by using a greater dispersion to open the slit of the spectroscope wide enough to see them, they have been made the subject of daily study and record. The results, so far as they have been investigated, show that the region of the Sun's limb in which the prominences are most frequent reaches to some 40° on either side of the equator, which is somewhat greater than the area of sun-spot frequency. The chromosphere itself is probably of much the same character as the erupted prominences, and formed of little flames arranged thickly together like "blades of grass."

In observing the Sun with a telescope the amateur will soon notice that the surface is far more brilliant in the central parts than round the margin of the disk. Vögel has estimated that immediately inside the edges the brightness does not amount to one seventh that of the centre. The difference is entirely due to the solar atmosphere, which is probably very shallow relatively to the great diameter of the Sun.

CHAPTER VI.

THE MOON.

Attractive aspect of the Moon.—Absence of air and water.—Only one
Hemisphere visible.—Earthshine.—Telescopic observations of the lunar
surface.—Eclipses.—Lunar changes.—Formations.—Plato and other objects described.—Table of Moon's age and formations near terminator.—
Occultations of stars.—Visibility of the new Moon.

> " The western Sun withdraws : meanwhile the Moon,
>   Full orb'd, and breaking through the scatter'd clouds,
>   Shows her broad visage in the crimson'd east."

EARLY in autumn, when the evenings are frequently clear,
many persons are led with more force than usual to evince an
interest in our satellite, and to desire information which may
not be conveniently obtained at the time.   The aspect of the
Moon at her rising, near the time of the full, during the
months of August, September, and October, is more conspicuously noticeable than at any other season of the year,
on account of the position she then assumes on successive
nights, enabling her to rise at closely identical times for
several evenings together.   The appearance of her large,
ruddy globe at near the same hour, and her increasing brilliancy as her horizontal rays give way under a more vertical
position, originated the title of " Harvest Moon," to comme-
morate the facility afforded by her light for the ingathering
of the corn preceding the time of the autumnal equinox.

It will be universally admitted that the Moon possesses
special attractions for us, as being situated nearer than any
other celestial body, and forming the inseparable companion or
tributary world to the Earth.   The many important influences
she exercises have led to her becoming the object of close investigation ; so that her motions and physical appearances have
been ascertained with a remarkable degree of exactness and

I

amplitude. Her movements regulate the tides; her positions are of the utmost moment to the mariner; her light is the welcome beacon of the wayfarer, and its picturesque serenity has ever formed the theme of poets. To the practical astronomer she constitutes an orb perfectly unique as regards extent and variety of detail; and questions relating to the physical condition of her surface, now and in past ages, supply a fund of endless speculation to the theorist.

The mean apparent diameter of the Moon is $31'$ $5''$, and it varies from $29'$ $21''$ at perigee to $33'$ $31''$ at apogee. Her real diameter is 2160 miles, and her mean distance slightly exceeds 237,000 miles. Her revolution round the Earth ($=$ sidereal period) is performed in 27^d 7^h 43^m $11^s{\cdot}46$, but the time from one new moon to another ($=$ synodical period) is 29^d 12^h 44^m 3^s. The Moon's motion through the firmament is at the rate of $13°$ $10'$ $35''$ per day and $32'$ $56''$ per hour. Thus she travels over a space slightly exceeding her own diameter in one hour. The linear value of $1''$ at the distance of our satellite is $1{\cdot}16$ mile, and of $1'$ $69\frac{1}{2}$ miles.

When we critically survey the face of the Moon with a good telescope, we see at once that her surface is broken up into a series of craters of various sizes, and that some very irregular formations are scattered here and there, which present a similar appearance to elevated mountain-ranges. The crateriform aspect of the Moon is perhaps the more striking feature, from its greater extent; and we recognize in the individual forms a *simile* to the circular cavities formed in slag or some other hard substances under the action of intense heat. In certain regions of the Moon, especially those near the south pole, the disk is one mass of abutting craters, and were it not for the obvious want of symmetry in form and uniformity of size, the appearance would be analogous to that of a gigantic honeycomb. These craters are commonly surrounded by high walls or ramparts, and often include conical hills rising from their centres to great heights. While the eye examines these singular structures, and lingers amongst the mass of intricate detail in which the whole surface abounds, we cannot but feel impressed at the marvellous sharpness of definition with which the different features are presented to our view. It matters not to what district we direct our gaze, there is the same perfect serenity and clearness of outline. Not the slightest indication can be discerned anywhere of mist or other obscuring vapours hanging over the lunar landscape.

Absence of Air and Water.—Now it is palpable from this that the Moon has no atmosphere of sufficient density to render itself appreciable; for such an appendage, if it existed in any visible form, would at once obtrude upon the attention, and we should probably recognize some of the characteristics common to the behaviour of our terrestrial clouds. But nothing of the kind is apparent on the Moon : there is an unbroken transparency spread over the whole extent of the Moon's scenery; whence we conclude that if any air exists on the surface it is of extremely attenuated nature, and possibly confined to the bottom of the craters and low-lying formations, which are arranged in such prolific manner on our satellite.

Nor is there any perceptible intimation of water upon the Moon. It is true that several dark grey patches have been given names, leading one directly to the inference that lakes and seas comprise part of the surface phenomena. Thus there is the *Mare Serenitatis* ("the sea of serenity") and many other designations of similar import, which we cannot but insist are wrongly applied and calculated to lead to mis-apprehension. Before the invention of the telescope furnished us with the means of accurately determining the character of the lunar features, such apellations may have been considered eligible ; but now that the non-existence of water in any extensive form is admitted, the titles are rendered obsolete. Still their retention is in some respects advisable, for any sweeping change in a recognized system of nomenclature must cause confusion, and the names alluded to serve a useful end in facilitating reference ; so that, under the circumstances, it would perhaps be unwise to attempt reform, or to introduce an innovation which must occasion many difficulties.

Only one Hemisphere visible.—In discussing the nature and appearances of the lunar formations, it must be dis-

tinctly understood that our remarks apply to those visible
on the side invariably turned towards the Earth. For, in
point of fact, there is a considerable expanse of the lunar
disk never perceptible from the Earth at all. This is occasioned by the circumstance that the Moon rotates upon her
axis in precisely the same time as she revolves around the
Earth, and is therefore enabled to present the same side
towards us on all occasions. A slight tilting (called libration)
takes place, so that we are allowed a glimpse of fragments of
the side normally invisible, and its analogous aspect leads us
to suppose that there is no great distinction between the
features of the Moon's visible and invisible hemispheres.
From exact computations it appears that we are enabled
to see a proportion of $\frac{59}{100}$ of the surface, and that the
remaining $\frac{41}{100}$ are permanently beyond our reach.

Earthshine.—A few mornings before new moon, and on a
few evenings after it, the whole outline of the dark portion
of the lunar globe may be distinctly perceived. A feeble illumination like twilight pervades the opaque part, and this is
really earthlight thrown upon our satellite, for near the
times of new moon the. Earth appears at her brightest (her
disk being fully illuminated) as seen from the Moon. The
French term for this light is *la lumière cendrée*, or "the
ashy light." The appearance is often popularly referred to in
our own country as "the old Moon in the new Moon's
arms." Some of the old observers remarked that the
waning Moon showed this earthlight more strongly than the
new Moon.

Telescopic Observations of the Lunar Surface.—Our telescopes give by far the most pleasing view of the Moon when
she is in a crescent shape. At such a period the craters and
mountains, with their dark shadows, are splendidly displayed.
A good view is also obtainable with the Moon at first or last
quarter, or when the disk is gibbous. But the full Moon is
decidedly less attractive ; for the shadows have all disappeared, and the various formations have quite lost their
distinctive character. The disk is enveloped in a flood of
light, causing glare, and though there is a large amount
of detail, including systems of bright rays, many differences

of tint, and bright spots, yet the effect is altogether less satisfactory than at the time of a crescent phase.

The nature of the work undertaken by the amateur must largely depend upon his opportunities and the capacity of his appliances. It is evident that in the investigation of lunar details it is essential to be very particular in recording observations ; for unless the conditions of illumination are nearly the same, lunar objects will present little resemblance. He should therefore examine the formations at intervals of 59^d 1^h 28^m, when the terminator is resting on nearly identical parts of the surface. In periods of 442^d 23^h ($=15$ lunations) there is another repetition of similar phase ; also in periods of 502^d 0^h 28^m ($=17$ lunations).

The observer, in entering results into his note-book, should state the Moon's age to the nearest minute, and give aperture and power of telescope and state of sky. Those objects which he has recorded at one lunation should be reobserved after an intervening lunation, or at intervals of 59^d 1^h 28^m. He will then find his notes and drawings are comparable. By the persistent scrutiny of special structures he will discern more and more of their details ; in other words, he will find his eye soon acquires power with experience and familiarity with the object. Comparisons of his own work with the charts and records of previous observers will be sure to interest him greatly, and the differences which he will almost certainly detect may exert a useful influence in inciting him to ascertain the source of them. He must not be premature in attributing such discordances to actual changes on the Moon ; for he must remember that perfect harmony is rarely to be found in the experiences of different observers. But whenever his own results are inconsistent with those of others, the fact should be carefully noted and the observations repeated and rediscussed with a view to reconcile them. The charts and descriptions of former selenographers are excellent in their way, and the outcome of much zealous labour ; but they contain omissions and inaccuracies which it has been impracticable to avoid. The amateur who discovers a mountain, craterlet, or rill not depicted on his lunar maps must therefore neither regard it as a new formation or as a new discovery ; for it may have been

overlooked by some of the previous observers, and is possibly
drawn or described in a work which he does not happen to
have consulted. Such differences should, however, always be
announced, as they clear the way for others working in the
same field.

A small instrument, with an object-glass of about $2\frac{1}{2}$ inches,
will reveal a large amount of intricate detail on the surface of
our satellite, and will afford the young student many evenings
of interesting recreation. But for a more advanced survey of
the formations, with the view to discover unknown objects or
traces of physical change in known features, a telescope of at
least 8 or 10 inches aperture is probably necessary, and powers
of 300, 350, and more.

Eclipses of the Moon.—These phenomena comprise a variety
of interesting aspects. They are less numerous, in actual
occurrence, than solar eclipses in the proportion of about
2 to 3; but they are more frequently visible, because they
may be witnessed from any part of an entire hemisphere,
whereas eclipses of the Sun are only observable from a tract
of the Earth's surface not exceeding 180 miles in breadth.
The Moon may remain totally eclipsed for a period of 2 hours
4 minutes, and the whole duration, including the penumbral
obscuration from its first to its last projection, is about 6 hours.
Sometimes the Moon suffers total eclipse twice in the same
year, and both may be visible, as in 1844, 1877, 1964, &c. It
is possible for three such eclipses to occur within a single year,
as in 1544. In 1917 there will be three total lunar eclipses,
but not all visible in England. In the latter year there will
be no less than seven eclipses, as in 1935.

On the last two occasions—Oct. 4, 1884, and Jan. 28,
1888—when the Moon was totally immersed in the Earth's
shadow, the atmosphere was very clear; and it is hoped
equally favourable conditions will attend the similar phenomena of Nov. 15, 1891, Sept. 4, 1895, and Dec. 27, 1898.
One of the most interesting features during these temporary
obscurations of our satellite is the occultation of small stars.
Prof. Struve compiled a list of no less than 116 of these objects
that would pass behind the Moon's shadowed limb during the
eclipse of Oct. 4, 1884.

Another important effect is the variable colouring on the Moon. This differs considerably in relative intensity as seen during successive eclipses, and the cause is not perhaps fully accounted for. Kepler thought it due to differences in humidity of those parts of the Earth's atmosphere through which the solar rays pass and are refracted to the eclipsed Moon. The intense red hue which envelopes the lunar surface on such occasions is due to the absorption of the blue rays of light by our atmosphere. The sky at sunset is often observed to be similarly coloured, and from the operation of similar causes. Sometimes the Moon entirely disappears when eclipsed, but on other occasions remains distinctly obvious, like a bright red ball suspended in the firmament. On May 5, 1110, Dec. 9, 1620, May 18, 1761, and June 10, 1816, our satellite is said to have become absolutely imperceptible during eclipse. Wargentin, who described the appearance in 1761, remarks :—" The Moon's body disappeared so completely that not the slightest trace of any portion of the lunar disk could be discerned, either with the naked eye or with the telescope." On Oct. 4, 1884, I noticed that the opacity was much greater than usual ; at the middle period of the eclipse the Moon's diameter was apparently so much reduced that she looked like a dull, faint, nebulous mass, without sharply determinate outlines. The effect was similar to that of a star or planet struggling through dense haze. Yet, on March 19, 1848, the Moon " presented a luminosity quite unusual. The light and dark places on the face of our satellite could be almost as well made out as on an ordinary dull moonlight night." On July 12, 1870, Feb. 27 and Aug. 23, 1877, and Jan. 28, 1888, the Moon, as observed at Bristol, was also fairly bright when totally immersed in the Earth's shadow. In explanation of these singular differences, Dr. Burder has suggested that Kepler's views seem inadequate, and that the solar corona is probably implicated in producing light and dark eclipses. He concludes that, as the corona sometimes extends to considerable distances from the Sun, and is very variable in brightness, it may have sufficient influence to occasion the effects alluded to.

Lunar Changes.—The question as to whether physical changes are occurring in the surface-formations of our satellite is one which offers attractive inducements to telescopic observers. Though the Moon appears to have passed the active state, it is very possible that trivial alterations continue to affect some of her features. In April 1787 Sir W. Herschel wrote :—"I perceive three volcanoes in different places of the dark part of the new Moon. Two of them are already nearly extinct, or otherwise in a state of going to break out ; the third shows an eruption of fire or luminous matter." Schröter, however, was correctly of opinion that these appearances were due to reflected light from the Earth falling upon elevated spots of the Moon having unusual capacity to return it. Schröter himself thought he detected sudden changes in 1791. He says that, on the 30th of December, at 5^h P.M., with a 7-foot reflector magnifying 161 times, he perceived the commencement of a small crater on the S.W. declivity of the volcanic mountain in the Mare Crisium, having a shadow of at least $2' \ 5''$. On the 11th of January, 1792, at $5^h \ 20^m$ P.M., on looking at the place again he could see neither the new crater nor its shadow. In this case the disappearance was doubtless an apparent one, merely due to the reversed illumination under which the object was examined in the interval of 12 days.

Many other observers besides Herschel have been struck with the brightness of certain spots situated in the opaque region of the lunar disk ; but there is no doubt the cause has been uniformly one and the same, viz. the highly reflective properties of some of the mountains (notably of one named Aristarchus), which are distinctly visible as luminous spots amid the relatively dark regions surrounding them. They afford no certain evidence of existing volcanic energy, and in the light of modern researches such an idea cannot be entertained.

On June 10, 1866, Temple noticed a remarkable light appearance, agreeing with the position of Aristarchus, upon the dark side of the Moon, faintly illuminated by earthshine. The object did not exhibit a faint white light analogous to

that of other craters in the dark side, but it was star-like, diffused, in colour reddish yellow, and evidently dissimilar to other bright spots. He wrote, in reference to this matter :—
" Of course I am far from surmising a still active chemical outbreak, as such an outbreak supposes water and an atmosphere, both of which are universally allowed not to exist on the Moon, so that the crater-forming process can only be thought of as a dry, chemical, although warm one."

On November 17, 1866, Schmidt announced that the lunar crater Linné, about 5½ miles in diameter, and situated in the Mare Serenitatis, had disappeared ! He averred that he had been familiar with the object as *a deep crater* since 1841, but in October 1866 he found its place occupied by *a whitish cloud.* This cloud was always visible, but the crater itself appeared to have become filled up, and was certainly invisible under its former aspect. Such a definite statement, emanating as it did from a diligent and experienced student of selenography, naturally aroused keen interest, and Linné at once became the object of wide-spread observation. But a reference to Schröter's results, obtained in the latter part of the last century, threw some doubt upon the alleged change. This observer had figured Linné on November 5, 1788, as a round white spot, and there is nothing in his drawing indicating a crateriform aspect. His description of Linné was :—" A flat, somewhat doubtful crater, which appears as a round white spot." Mr. Huggins regarded Schröter's observations as correctly expressing the appearance of this object in 1867 under the same conditions of illumination. On the other hand, Lohrmann (1823) and Mädler (1831) referred to Linné as a deep crater, and in terms inconsistent both with Schröter's drawing and with the present aspect of the object. The outcome of the many fresh observations that were collected was that Linné appeared as a white cloud, with a small black crater within a large shallow-ringed depression. But as usual in such cases, the observers were far from being unanimous as to the details of the formation; and certainly in regard to a lunar object this need occasion no surprise, for slight differences in the angle of illumination produce marked changes in the aspect of lunar features. The fact of actual change could not be demonstrated, and the negative view appears to have subsequently gained weight.

Another instance of alleged activity on the Moon was notified by Dr. Klein in the spring of 1877. He saw a deep black crater about 18 miles to the W.N.W. of Hyginus, and in a particular place where he had previously recognized no such object, though he had frequently examined the region and was perfectly familiar with it. Forthwith every telescope was directed to this part of the Moon. The maps of earlier observers were eagerly consulted, and lunar photographs scanned for traces of the new object. Many drawings were made of the district near Hyginus and of the remarkable rill or cleft connected with it ; but amongst both old and new records some puzzling discordances were detected. Many of the observers, instead of finding Dr. Klein's new formation a sharply-cut, deep crater, saw it rather in the character of a saucer-like depression ; and I drew it under this aspect on several occasions with a 10-inch reflector. The fact, therefore, of its being a new feature admitted of no valid and convincing proofs, and thus the same uncertainty remains attached to this object as to Linné, nothing being absolutely proved *. The problem as to whether the Moon is still the seat of physical activity has yet to be solved.

Many circumstances are antagonistic to the discovery of changes on the Moon. As the Sun's altitude is constantly varying with reference to lunar objects, they assume different aspects from hour to hour. In a short interval the same formations become very dissimilar. When the Sun is rising above the more minute craters they are often distinguished in their true characters ; but near the period of full Moon they are visible as bright spots, and it is impossible to tell whether they represent craters or conical hills. With a vertical Sun, as at the full, all the shadows have disappeared—in fact, the entire configuration has been transformed, and many of the

---

* In September 1889 Prof. Thury, of Geneva, reported a change in the centre of the crater Plinius. With a 6-inch refractor he saw, instead of the usual two hills in the interior, a circular chalk-like disk " with a dark spot in its centre like the orifice of a mud-volcano."

interesting lineaments displayed at the crescent phase are no longer seen. The Moon's libration also introduces slight differences in the appearance of objects. And these are not the only drawbacks ; for observations, in themselves, are seldom accordant, and it is found that drawings and descriptions are not always to be reconciled, though referring to identical and invariable features. The lunar landscape must be studied under the same conditions of illumination and libration, with the same instrument and power, and in a similar state of atmosphere, if results are to be strictly comparable. But it is very rarely that observations can be effected under precisely equal conditions ; hence discordances are found amongst the records.

The whole of the Moon's visible sphere exhibits striking imprints of convulsions and volcanic action in past times, though no such forces appear to operate now. The surface seems to have become quiescent, and to have assumed a rigidity inconsistent with the idea of present energy. But we cannot be absolutely certain that minute changes are not taking place, and, being minute, the prospect of their detection is somewhat remote. Students of lunar scenery will probably have to watch details with scrupulous care and for long periods before an instance of real activity can be demonstrated.

Lunar Formations.—The Moon abounds in objects of very diversified character, and they have been classified according to peculiarities of structure. The names of eminent astronomers have been applied to many of the more definite features—a plan of nomenclature which originated with Riccioli, who published a lunar map at the middle of the seventeenth century. The following brief summary comprises many of the principal formations :—

Mare. A name applied by Hevelius to denote the large and relatively level plains on the Moon, which present some similarity to terrestrial seas. They are visible to the naked eye as dusky spots, and in a telescope show many craters, hills, and mounds, and some extensive undulations of surface.

Palus (Marsh) and *Lacus* (Lake) were titles given by Riccioli to minor areas of a dark colour, and exhibiting greater variety of detail and tint than the *Maria*.

Sinus (Bay) has been applied to objects like deep bays on the borders of the *Maria.*

Walled Plains extend from 40 to 150 miles in diameter, and are commonly surrounded by a terraced wall or mountain-ranges. The interiors are tolerably level, though often marked with crater-pits, mounds, and ridges.

Mountain-Rings. These represent rings of mountains and hills, enclosing irregularities, possibly furnished by the debris of the crumbling exterior walls, which, in certain instances, appear to have fallen inwards.

Ring-Plains are more circular and regular in type than the walled plains, and consist of a moderately flat surface surrounded by a single wall. *Crater-Plains* are somewhat similar, and seldom exceed 20 miles in diameter. They " rise steeply from the mass of debris around the foot of their walls to a considerable height, and then fall precipitously to the interior in a rough curved slope, whilst on their walls, especially on the exterior, craterlets and crater-cones often exist in considerable numbers."

Craters, Craterlets, and Crater-Pits. Usually circular in form, and severally offering distinctions as to dimensions and shape. The craters are surrounded by walls, rising abruptly to tolerable heights, and pretty regular in their contour. When the Sun is rising the shadow of the walls falls upon the interior of the craters, and many of these dark conspicuous objects are to be seen near the Moon's terminator. With a high Sun some of the craters are extremely bright. In proof of the large number of these objects, it may be noted here that in Mädler's lunar map (1837) 7735 craters are figured, while in Schmidt's (1878) there are no less than 32,856 !

Crater-Cones. Conical hills or mountains, visible as small luminous spots about the period of full Moon. They are from ½ to 3 miles in diameter, and show deep central depressions. It is somewhat difficult to distinguish them from the ordinary mountain-peaks and white spots, and they are not unlike the cones of terrestrial volcanoes.

Rills or Clefts. These are very curious objects. They were first discovered by Schröter in 1787, and some of them are to be traced over a considerable extent of the lunar surface, their entire length being 200 or 300 miles. They have the appearance of cuttings or canals, and are sometimes straight, sometimes bent, and not unfrequently develop branches which intersect each other. They apparently run without interruption through many varieties of lunar objects. The bottoms of these rills are nearly flat, and look not unlike dried river-beds. Some observers have regarded them as fractures or cracks in the Moon's surface ; but their appearance and circumstances of arrangement are opposed to such a view. Our present knowledge includes more than 1000 of these rills.

Mountain-Ranges are chains of lofty peaks and highlands, sometimes divided by rills and numerous ravines and cross valleys. Some of these ranges are of vast magnitude, and the summits of the mountains reach altitudes between 15,000 and 20,000 feet, and sometimes even more.

Mountain-Ridges are to be found scattered in the greatest abundance in the most disturbed localities of the lunar surface. They sometimes connect several formations, or surmount ravines or depressions of large extent. Peaks attaining altitudes of more than 5000 feet rise from them, and they range in several cases over 100 miles.

Ray-Centres. Systems of radiating light-streaks, having a mountain-ring as the centre of divergence, and stretching to distances of some hundreds of miles round. Tycho, Copernicus, Kepler, Anaxagoras, Aristarchus, and Olbers are pronounced examples of this class.

In Beer and Mädler's chart of the Moon the names are attached to the various formations, as they are also in Neison's maps and in some other works. One of these will be absolutely necessary to the student in prosecuting his studies. He will then have a ready means of acquainting himself with the various formations, and making comparisons between his new results and the drawings of earlier selenographers. I would refer the reader to Neison's and Webb's books for many references in detail to lunar features, and must be content here with a brief description of a few leading objects :—

Plato is an extensive walled plain, 60 miles in diameter, and situated on the N.E. boundary of the Mare Imbrium.

Nasmyth and Carpenter describe the wall as "serrated with noble peaks, which cast their black shadows across the plateau in a most picturesque manner, like the towers and spires of a great cathedral." It has received a large amount of attention, with a view to trace whether changes are occurring in the numerous white spots and streaks lying in its interior. In 1869–71 Mr. Birt collected many observations, and on discussing them was led to believe that "there is strong probability that activity, of a character sufficient to render its

Fig. 25.

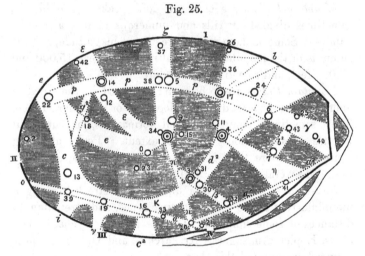

Light-spots and streaks on Plato, 1879-82.   (A. Stanley Williams.)

effects visible, had been manifested." The inquiry was renewed by Stanley Williams in 1882–84, and he concluded that the results were strongly confirmatory of actual change having occurred since 1869–71. The relative visibility of several of the bright spots had altered in the interim, and the curious intermingling bright streaks also exhibited traces of variation. At sunrise the interior of Plato is pure grey; but with the sun at a considerable height above it, the plain becomes a dark steel-grey. The change is an abnormal one, and difficult to explain. South of Plato there is a fine example of an isolated peak, named Pico, which is about 8000 feet high.

Great Alpine Valley. This object, supposed to have been discovered by Bianchini in 1727, and having a length, according to Mädler, of 83 miles and a breadth varying from $3\frac{1}{2}$ to 6 miles, is a very conspicuous depression situated near Plato, and running from the Mare Frigoris to the Mare Imbrium. It exhibits at its southern extremity an oval formation, and a narrow gorge issues from it to the northward, opening out further on, and imparting to the whole appearance a shape which Webb likened to a Florence oil-flask. Elger has fully described this singular structure. " It is only when far removed from the terminator that its V-shaped outlet to the Mare Imbrium flanked on either side by the lofty Alps can be traced to advantage, or the flask-like expansion with the constricted gorge leading up to it from the N.W. satisfactorily observed. At other times these features are always more or less concealed by the shadows of neighbouring heights. The details of the upper or more attenuated end of the valley are, however, best seen under a setting sun, when many striking objects come to light, of which few traces appear at other times."

Archimedes. One of the most definite and regular of the walled plains. It is 60 miles in diameter, with a wall rising about 4200 feet above the surface. Some small craters and various streaks diversify its centre.

Tycho. A grand ring-plain, 54 miles in diameter and about 17,000 feet (=nearly 3 miles) deep, and forming the centre of the chief ray-system of the Moon. The light-radiations stretch over one fourth of the visible hemisphere at the full, but they are imperceptible with the Sun's altitude below 20°. These remarkable radiations from Tycho form a striking aspect of lunar scenery, and any small telescope reveals them. Webb has termed Tycho " the metropolitan formation of the Moon ;" and the idea embodied in the expression must strike observers as very apposite. This object is visible to the naked eye at the time of full. A fine hill rises from its centre to a height of 5500 feet.

Copernicus. A magnificent ring-plain, 56 miles in diameter, and surrounded by a wall (in which there are terraces and lofty peaks, separated by ravines) attaining an elevation of 11,000 feet. There is a central hill of nearly 2500 feet. From Copernicus light-streaks are plentifully extended on all sides, and apparently connect this object with the many others of similar character which are situated in this region. Neison says that near Copernicus the light-streaks unite and form a kind of nimbus or light-cloud about it. The streaks are most conspicuous towards the N., where they are from 5 to 14 miles in width. To the N.W. of Copernicus, about halfway in the direction of the neighbouring ring-plain *Eratosthenes* (and N. of Stadius), there is a considerable number of crater-pits. Mädler figured sixty-one of these, and regarded that number as certainly less than half the total number visible. They appear to be ranged in rows or streams, and are so close together in places as to nearly form crater-rills. Schmidt saw the ground hereabout pierced like a honeycomb, and managed to count about 300 little craters ; but they are so thickly strewn in this district that exact numbers or places cannot be assigned. They are best observable when the Sun is rising on the E. wall of Copernicus. The interior of this fine object shows six or seven peaks, which are often capped with sunshine, and very brilliant amid the black shadow thrown from the surrounding wall.

Theophilus. Another ring-plain, and one of the deepest visible. Its terraced lofty wall, 64 miles in diameter, rises in a series of peaks to heights varying between 14,000 and 18,000 feet. There is a central mountain, broken by ravines; but from one of the masses a peak ascends to a height of about 6000 feet.

Petavius. A large walled plain, surrounded by a double wall or rampart, which rises to 11,000 feet on its E. side. There are hills and ridges in the interior, and a central peak, A, reaching to 5500 feet above the E. part of the floor, which is convex in form. A smaller peak, of nearly 4000 feet, lies W. of A. Several small craterlets have been seen in the interior.

Newton. The deepest walled plain known upon the Moon's surface. In form it is elliptical ; its length is 143 miles, while its breadth is only 69 miles. The walls show the terracing so common in these objects, and one lofty peak reaches the unusual height of 24,000 feet above the floor. The interior includes some small craters, mountain protuberances, and other irregularities. Neison says that, owing

Fig. 26.

Petavius and Wrottesley at sunset.   1885, Dec. 23, 9ʰ to 10ʰ 30ᵐ.
(T. Gwyn Elger.)

to " the immense height of the wall, a great part of the floor is entirely lost in shadow, neither Earth nor Sun being ever visible from it."

Grimaldi. An immense walled plain, extending over 148 miles from N. to S. and about 130 miles from E. to W. Its interior is very dark. *Clavius* is another grand example

of this class of object, and is rather larger than Grimaldi, but
unfavourably placed near the S. pole.　*Schickard* may also be
mentioned as a large formation of similar type, and situated
near the S.E. limb of the Moon.

Fig. 27.

Birt, Birt A, and the Straight Wall.　1883, Feb. 15, 6ʰ to 8ʰ 40ᵐ,
(T. Gwyn Elger.)

Rill or Cleft of Hyginus. A conspicuous example of the
lunar rills, and one which yields to very moderate instruments.　Neison notes that it is readily visible in a 2-inch
telescope ; while Webb remarks that a power of only 40, in a
good instrument, is enough to show it under any illumination.
The rill is about 150 miles long.　It cuts through a number

of crater-pits, and Mädler found so many widenings in it that it appeared like a confluent train of craters. The rill traverses the large crater-pit Hyginus, which is $3\frac{3}{4}$ miles in diameter and moderately deep. Other fine examples of rill-systems will be found between Rheita and Metius and near Triesnecker and Ramsden.

Straight Wall. A singular structure on the E. side of the ring-plain Thebit. It is a ridge or wall, which looks regular enough for a work of art, according to Webb. Its average height is 450 feet (Schröter), 1004 feet (Mädler), or 880 feet (Schmidt). These several determinations are given to show the discordances sometimes found in the measures of good observers. This object is about 60 miles long ; at one extremity lies a small crater, at the other there is a branching mountain nearly 2000 feet high. Elger has drawn this object, under both a rising and a setting sun, in the Liverpool Astronomical Society's ' Journal,' vol. v. p. 156, and remarks that it may be well observed at from 20 to 30 hours after the Moon's first quarter.

Valley near Rheita. South of the ring-plain Rheita, on the S.W. limb, there is an enormous valley, which extends in its entire length over 187 miles, with a width ranging from 10 to 25 miles. There are several fine valleys in this particular region.

Leibnitz Mountains. These are really situated on the further hemisphere of the Moon, but libration brings them into view, and they are sometimes grandly seen in profile on the S. margin. Four of the peaks ascend to elevations of 26,000 or 27,000 feet, and one mass, towering far above the others, is fully 30,000 feet in height, and is unquestionably the most lofty mountain on the Moon.

Dörfel Mountains. Visible on the Moon's S.S.E. limb. They exhibit three peaks, which, on the authority of Schröter, rise to more than 26,000 feet above the average level of the limb. The loftiest mountains on the Earth are in the Himalayas—a range of immense extent to the N. of India. The three highest peaks are Mount Everest (29,002 feet), Kunchinjinga (28,156 feet), and Dhawalagiri (28,000 feet). The only lunar mountain more elevated than these is that of the Leibnitz range, which, as we have already stated, ascends to fully 30,000 feet.

Apennines. A vast chain of mountains, extending over more than 450 miles of the lunar surface. *Huygens* is

Fig. 28.

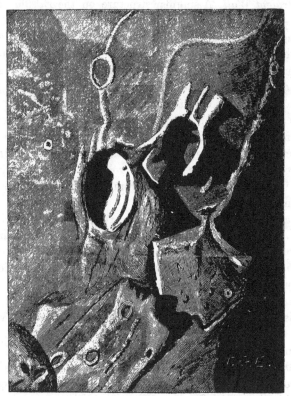

Aristarchus and Herodotus at sunrise. 1884, Jan. 9, 8ʰ 30ᵐ to 10ʰ 30ᵐ.
(T. Gwyn Elger.)

the most elevated peak, rising to more than 18,000 feet, and on its summit it shows a small crater. There are several other very lofty peaks in this range. The Sun rises upon the westerly region of these mountains at the time of first quarter, and the peaks and ridges, with their contrasting shadows, create a gorgeous effect just within, and projecting into the darkness beyond, the terminator. There is an immense amount of detail to be studied here, and much of it is within the reach of small instruments.

As the lunar mountains and craters are best seen near the terminator, it may be useful to give a table of objects thus favourably placed between the times of new and full Moon. The summary may assist the student, though it does not aim at exactness, only even days being given.

Moon's age in days.

Objects near the Terminator.

2 Mare Crisium, Messala, Sunrise on the Mare Humboldtianum, Langrenus, Vendelinus, Condorcet, Hansen, Gauss[β], Hahn, Berosus.

3[α] Craters in Mare Crisium, Taruntius, Picard, Fraunhofer, Vega, Pontécoulant, Cleomedes[γ], Furnerius, Petavius, Endymion, Messier[δ], Vlacq.

4 Mare Nectaris, Macrobius[ε], Proclus, Sunrise on Fracastorius, Rheita and Metius with the intervening valley, Guttemberg, Colombo, Santbech, Mountainous region W. of Mare Serenitatis, Hercules, Atlas.

5 Palus Somnii, Plana, Capella, Isidorus, Polybius, Piccolomini, Vitruvius, Littrow, Fabricius, Posidonius, LeMonnier, Theophilus, Cyrillus, Catharina, Hommel.

6 Tacitus, Maurolycus, Barocius, Dionysius[ζ], Sosigenes, Abulfeda, Descartes, Almamon, Gemma Frisius, Plinius, Ross, Arago, Delambre, Aristoteles, Eudoxus, Julius Cæsar, Linné, Menelaus.

[α] The objects for observation when the Moon's age is from 2 to 4 days may be suitably re-examined a few days after the full.

[β] An extensive walled plain, 110 miles in length.

[γ] A large walled plain containing a small crater, Cleomedes A.

[δ] A curious double crater, with comet-like rays crossing the Mare Fœcunditatis.

[ε] A circular ring-plain, 42 miles in diameter.

[ζ] The interior of this crater exhibits some interesting features as the Sun rises higher above it.

Moon's age.

days.

7 Ptolemæus, Albategnius, Manilius[η], Hyginus and its rill-system, Hipparchus, Autolycus, Aristillus, Cassini, Alpine Valley, W. C. Bond, Walter, Miller, LaCaille, Apennines, Triesnecker and the rills W. of it.

8 Mare Frigoris, Arzachel, Alphonsus, Alpetragius, Bode, Pallas, Archimedes, Plato, Maginus[θ], Mösting[ι], Thebit, Saussure, Moretus, Straight Wall, Lalande, Kirch.

9 Tycho, Clavius, Eratosthenes[κ], Stadius and the craters running to N.E., Timocharis, Pitatus, Gruemberger, Teneriffe Mountains, Straight Range[λ], Formation W. of Fontenelle[μ], Gambart.

10 Sinus Iridum, Copernicus, Hesiodus and the rill to E., Wilhelm I., Longomontanus[ν], Heinsius, Pytheas, Lambert, Helicon, Wurzelbauer.

11 Bullialdus, Campanus, Mercator, Reinhold, Riphæan Mountains, Hippalus, Capuanus, Blancanus, Tobias Mayer.

12 Mare Imbrium, Gassendi[ξ], Aristarchus and sinuous valley to the N.E., Herodotus, Marius, Flamsteed, Letronne, Schiller, Mersenius, Doppelmayer.

13 Schickhard, Wargentin, Grimaldi, Byrgius, Phocylides, Hevelius, Seleucus, Crüger, Briggs, Segner, Sirsalis.

[η] A fine ring-plain, 25½ miles in diameter.

[θ] Mädler says "the full Moon knows no Maginus,' meaning that this object is invisible under a vertical Sun.

[ι] Mösting, Lalande, and Herschel form a fine triangle when the Su has attained a great altitude. Mösting is a ray-centre.

[κ] A ring-plain 37½ miles in diameter, with very irregular terraced walls.

[λ] A range of mountains, with intervening valleys.

[μ] Mädler describes this as a square enclosure with rampart-like boundaries, which "throw the observer into the highest astonishment."

[ν] A great walled plain, 91 miles in diameter.

[ξ] A walled plain, 55 miles in diameter, in which Schröter suspected changes.

Moon's age.
days.

14   Mare Smythii, Bailly, Inghirami, Bouvard, Riccioli,
      Olbers, Hercynian Mountains, Cardanus, Krafft,
      Cordilleras[o], Pythagoras[π].

Occultations of Stars.—Among the various phenomena to
which the lunar motions give rise none are more pleasing to
the possessors of small telescopes than occultations of stars.
Several of these occurrences are visible every month.   If the
amateur has the means of obtaining accurate time, he will engage himself usefully in noting the moments of disappearance
and reappearance of the stars occulted.   This work is efficiently
done, it is true, at some of our observatories, and therefore little
real necessity exists for amateurs to embark in routine work
which can be conveniently undertaken at establishments where
they have better appliances and trained observers to use them.
The mere watching of an occultation, apart from the registry of
exact results, is interesting ; and there are features connected
with it which have proved exceedingly difficult to account for.
The stars do not always disappear instantaneously.   On coming
up to the edge of the Moon they have not been suddenly
blotted out, but have appeared to hang on the Moon's limb for
several seconds.   This must arise from an optical illusion, from
the action of a lunar atmosphere, or the stars must be observed
through fissures on the Moon's edge.   The former explanation
is probably correct ; for it has happened that two observers at
the same place have received different impressions of the phenomenon.   One has seen the star apparently projected on the
Moon's limb for about 5 seconds, while the other has witnessed
its sudden extinction, in the usual manner, as it met the Moon's
edge.   New observations, made with good instruments and
reliable eyes, and fully described, will doubtless throw more
light on the peculiar effects sometimes recorded.

   [o] An extensive mountain-range on the E. by S. limb.

   [π] A walled plain, 95 miles in diameter, and probably the deepest in
the N.E. quadrant, for the S.E. side of its wall rises to nearly 17,000 feet.
   After the full the same objects should be re-examined under the
reversed illumination.

Visibility of the new and old Moon.—It is an interesting feature of observation to note how soon after conjunction the Moon's thin crescent is observable with the naked eye. A case has been mentioned in which the old Moon was seen one morning before sunrise and the new Moon just after sunset on the next day. At Bristol, on the evening of March 30, 1881, I saw the new Moon at 7^h 10^m, the horizon being very clear in the west. She was then only 20^h 38^m old. On June 4, 1875, I observed the Moon's crescent at 9^h 10^m, or 22^h 49^m after new Moon. Dr. Degroupet, of Belgium, saw the old Moon on the morning of Nov. 22, 1889, between 6^h 47^m and 7^h 22^m G.M.T., or within 18^h 22^m of the time of new Moon.

CHAPTER VII.

MERCURY.

Supposed planet, "Vulcan."—Visibility of Mercury.—Period &c.—
Elongations.— Amateur's first view.— Phases. — Atmosphere of Mer-
cury.—Telescopic observations.—Schiaparelli's results.—Observations of
Schröter and Sir W. Herschel.—Transits of Mercury.—Occultations of
Mercury.

> " Come, let us view the glowing west,
>   Not far from the fallen Sun ;
> For Mercury is sparkling there,
>   And his race will soon be run.
> With aspect pale, and wav'ring beam,
>   He is quick to steal away,
> And veils his face in curling mists,—
>   Let us watch him while we may."

Supposed planet "Vulcan."—Mercury is the nearest known
planet to the Sun. It is true that a body, provisionally
named Vulcan*, has been presumed to exist in the space
interior to the orbit of Mercury ; but absolute proof is
lacking, and every year the idea is losing strength in the
absence of any confirmation of a reliable kind. Certain
planetary spots, observed in motion on the solar disk, were
reported to have been transits of this intra-Mercurial orb.
Some eminent astronomers were thus drawn to take an
affirmative view of the question, and went so far as to
compute the orbital elements and predict a few ensuing
transits of the suspected planet. But nothing was seen
at the important times, and some of the earlier observations
have been shown to possess no significance whatever, while
grave doubts are attached to many of the others. Not one
of the regular and best observers of the Sun has recently

---

* Chambers, in his ' Descriptive Astronomy,' 4th edition, 1889, devotes
a chapter to the discussion of facts having reference to Vulcan ; and the
reader desiring full information will find it here.

detected any such body during its transits (which would be likely to occur pretty frequently), and there is other evidence of a negative character; so that the ghost of Vulcan may be said to have been laid, and we may regard it as proven that no major planet revolves in the interval of 36,000,000 miles separating Mercury from the Sun.

Visibility of Mercury.—Copernicus, amid the fogs of the Vistula, looked for Mercury in vain, and complained in his last hours that he had never seen it. Tycho Brahe, in the Island of Hueen, appears to have been far more successful. The planet is extremely fugitive in his appearances, but is not nearly so difficult to find as many suppose. Whenever the horizon is very clear, and the planet well placed, a small sparkling object, looking more like a scintillating star than a planetary body, will be detected at a low altitude and may be followed to the horizon.

Period &c. — Mercury revolves round the Sun in 87ᵈ 23ʰ 15ᵐ 44ˢ in an eccentric orbit, so that his distance from that luminary varies from 43,350,000 to 28,570,000 miles. When in superior conjunction the apparent diameter of the planet is 4″·5; at inferior conjunction it is 12″·9, and at elongation 7″. His real diameter is 3000 miles.

Elongations.—Being situated so near to the Sun, it is obvious that to an observer on the Earth he must always remain in the same general region of the firmament as that body. His orbital motion enables him to successively assume positions to the E. and W. of the Sun, and these are known as his elongations, which vary in distance from 18°. to 28°. He becomes visible at these periods either in the morning or evening twilight, and under the best circumstances may remain above the horizon two hours in the absence of the Sun. The best times to observe the planet are at his E. elongations during the first half of the year, or at his W. elongations in the last half; for his position at such times being N. of the Sun's place, he remains a long while in view. It is unfortunate that when the elongation approaches its extreme limits of 28° the planet is situated S. of the Sun, and therefore not nearly so favourably visible as at an elongation of only 18° or 20°, when his position is N. of the Sun.

I have seen Mercury on about sixty-five occasions with the naked eye. In May 1876 I noticed the planet on thirteen different evenings, and between April 22 and May 11, 1890, I succeeded on ten evenings. I believe that anyone who made it a practice to obtain naked-eye views of this object would succeed from about twelve to fifteen times in a year. In a finer climate, of course, Mercury may be distinguished more frequently. Occasionally he presents quite a conspicuous aspect on the horizon, as in February 1868, when I thought his lustre vied with that of Jupiter, and in November 1882, when he shone brighter than Sirius. The planet is generally most conspicuous *a few mornings after his W. elongations* and *a few evenings before his E. elongations.*

Amateur's First View.—The first view of Mercury forms quite an event in the experience of many amateurs. The evasive planet is sought for with the same keen enthusiasm as though an important discovery were involved. For a few evenings efforts are vain, until at length a clearer sky and a closer watch enables the glittering little stranger to be caught amid the vapours of the horizon. The observer is delighted, and, proud of his success, he forthwith calls out the members of his family that they, too, may have a glimpse of the fugitive orb never seen by the eye of Copernicus.

Phases.—In the course of his orbital round Mercury exhibits all the phases of the Moon. Near his elongations the disk is about half illuminated, and similar in form to that of our satellite when in the first or third quarter. But the phase is not to be distinctly made out unless circumstances are propitious. Galilei's telescope failed to reveal it, and Hevelius, many years afterwards, found it difficult. This is explained by the small diameter of the planet and the rarity with which his disk appears sharply defined. The phase is sometimes noted to be less than theory indicates ; for the planet has been seen crescented when he should have presented the form of a semicircle. Several observers have also remarked that his surface displays a rosy tint, and that the terminator is more deeply shaded and indefinite than that of Venus.

Atmosphere.—The atmosphere of Mercury is probably far less dense than that of Venus. The latter being farthest from the

Sun might be expected to shine relatively more faintly than the former, but the reverse is the case. Mercury has a dingy aspect in comparison with the bright white lustre of Venus. On May 12, 1890, when the two planets were visible as evening stars, and separated from each other by a distance of only 2°, I examined them in a 10-inch reflector, power 145. The disk of Venus looked like newly-polished silver, while that of Mercury appeared of a dull leaden hue. A similar observation was made by Mr. Nasmyth on September 28, 1878. The explanation appears to be that the atmosphere of Mercury is of great rarity, and incapable of reflection in the same high degree as the dense atmosphere of Venus.

Telescopic Observations.—As this planet is comparatively seldom to be observed under satisfactory conditions, it is scarcely surprising that our knowledge of his appearance is very meagre, or that amateurs consider the planet an object practically inaccessible as regards the observation of physical peculiarities, and one upon which it is utterly useless to apply the telescope in the hope of effecting new discoveries. Former attempts have proved the extreme difficulty of obtaining good images of this planet. The smallness of the disk, and the fact that it is usually so much affected by the waves of vapour passing along the horizon as to be constantly flaring and moulding in a manner which scarcely enables the phase to be made out, are great drawbacks, which render it impossible to distinguish any delicate features that may be presented on the surface.

These circumstances are well calculated to lead observers to abandon this object as one too unpromising for further study; but I think the view is partly induced by a misconception. The planet's diminutive size is a hindrance which cannot be overcome ; but the bad definition, resulting from low altitude, may be obviated by those who will select more suitable times for their observations and not be dismayed if their initiatory efforts prove futile. As a naked-eye object, Mercury must necessarily be looked for when near the horizon ; but there is no such need in regard to telescopic observation, which ought to be only attempted when the planet surmounts the dense lower vapours and is placed at a sufficient elevation to give the instrument a fair chance of producing a steady image. The presence of sunshine need not seriously impair the definition or make the disk too faint for detail.

I have occasionally seen Mercury, about two or three hours after his rising, with outlines of extreme sharpness and quite comparable with the excellent views obtained of Venus at the time of sunrise or sunset. Those who possess equatoreals should pick up the planet in the afternoon and follow him until after sunset, when the horizontal vapours will interfere. Others who work with ordinary alt-azimuth stands will find it best to examine the planet at his western elongations during the last half of the year, when he may be found soon after rising by the naked eye or with an opera-glass, and retained in the telescope for several hours after sunrise if necessary. He may sometimes also be brought into the field before sunset (at the eastern elongations in the spring months), by careful sweeping with a comet-eyepiece, especially when either the Moon, Venus, or Jupiter happens to be near, and the observer has found the relative place of the planet from an ephemeris.

Schiaparelli's Results.—Mercury was displayed under several advantages in the morning twilight of November 1882, and I made a series of observations with a 10-inch reflector, power 212. Several dark markings were perceived, and a conspicuous white spot. The general appearance of the disk was similar to that of Mars, and I forwarded a summary of my results to Prof. Schiaparelli, of Milan, who favoured me with the following interesting reply:—

" I have myself been occupied with this planet during the past year (1882). You are right in saying that Mercury is much easier to observe than Venus, and that his aspect resembles Mars more than any other of the planets of the solar system. It has some spots which become partially obscured and sometimes completely so; it has also some brilliant white spots in a variable position. As I observe the planet entirely by day and near the meridian I have been able to see its spots many times, but not always with the necessary distinctness to make drawings sufficiently reliable to serve as a base for a rigorous investigation. It is remarkable that the views taken near superior conjunction have been more instructive for me than those taken when the disk is near dichotomy, the defect in diameter being compensated by the possibility of seeing nearly all the disk, which, under those conditions, is more strongly illuminated. I believe that by instrumental means, such as our 8½-inch refractor at Milan gives, it is possible to prove the rotation-period of Mercury and to gain a knowledge of the principal spots as regards the generality of their forms. But these spots are really very complicated, for, besides the difficulties attending their observation, they are extremely variable."

Prof. Schiaparelli used an 8½-inch refractor in this work, and was able, under some favourable conditions, to apply a power of 400. The outcome of his researches, encouraged since 1882 by the addition of an 18-inch refractor to the appliances of his Observatory, has been recently announced in the curious fact that the rotation of Mercury is performed in the same time that the planet revolves round the Sun ! If this conclusion is just, Mercury constantly presents one and the same hemisphere to the Sun, and the behaviour of the Moon relatively to the Earth has found an analogy. But these deductions of the eminent Italian astronomer require corroboration, and this is not likely to be soon forthcoming owing to the obstacles which stand in the way.

Observations of Schröter and Sir W. Herschel.—Schröter observed Mercury with characteristic diligence between 1780 and 1815. In 1800 he several times remarked that the southern horn of the crescent was blunted, and fixed the planet's rotation-period at 24^h 4^m. He also inferred the existence of a mountain 12 miles in height. But elements of doubt are attached to some of Schröter's observations; and Sir W. Herschel, whose telescopic surveys of both Mercury and Venus were singularly barren of interesting results, pointed out their improbability. But the great observer of Slough was not very amicably disposed towards his rival in Germany. His strictures appear, however, to have been not without justice if we consider them in the light of modern observations.

Surface-markings.—Spots or markings of any kind have rarely been distinguished on Mercury. On June 11, 1867,

Prince recorded a bright spot, with faint lines diverging from
it N.E. and S.   The spot was a little S. of the centre.   Birmingham, on March 13, 1870, glimpsed a large white spot near
the planet's E. limb, and Vögel, at Bothkamp, observed spots
on April 14 and 22, 1871.   These instances are quoted by
Webb, and they, in combination with the markings seen by

Fig. 29.

1882, Nov. 5, 18ʰ 49ᵐ.   1882, Nov. 6, 18ʰ 55ᵐ.

Mercury as a morning star.   (10-inch Reflector; power 212.)

Schiaparelli at Milan and by the author at Bristol in 1882,
sufficiently attest that this object deserves more attentive
study.

Amateurs with moderately large instruments would be
usefully employed in following this planet at the most
opportune periods and making careful drawings under the
highest powers that can be successfully applied.   Mercury
has been persistently neglected by many in past years, and no
doubt this " swift-winged messenger of the Gods" has eluded
some of his would-be pursuers ; but there is every prospect that
a patient observer, careful to utilize all available opportunities,
would soon gather some profitable data relating to his
appearance.

Transits of Mercury.—One of the most interesting phenomena, albeit a somewhat rare event, in connection with

Mercury, is that of a transit across the Sun. The planet then appears as a black circular spot. Observers have noticed one or two very small luminous points on the black disk, and an annulus has been visible round it. These features are probably optical effects, and it will be worth while to remember them on the occasion of future transits, of which the subjoined is a list :—

1891, May 9.	1937, May 10.
1894, Nov. 10.	1940, Nov. 12.
1907, Nov. 12.	1953, Nov. 13.
1914, Nov. 6.	1960, Nov. 6.
1924, May 7.	1970, May 9.
1927, Nov. 8.	1973, Nov. 9.

The first observer of a transit of Mercury appears to have been Gassendi, at Paris, on Nov. 7, 1631.

Occultations of Mercury.—There was an occultation of Mercury by the Moon on April 25, 1838. It occurred on the day of the planet's greatest elongation E., and at a time in the evening when it might have been most suitably witnessed, but cloudy skies appear to have frustrated the hopes of intending observers. There was a repetition of the event on the morning of May 2, 1867, and it occurred, curiously enough, less than 24 hours after an occultation of Venus.

CHAPTER VIII.

VENUS.

Beauty of Venus.—Brilliancy.—Period &c.—Venus as a telescopic object.—Surface-markings on the planet.—Rotation-period.—Faintness of the markings.—Twilight on Venus.—Alleged Satellite.—Further observations required.—Transits of Venus.—Occultations of Venus.

> " Friend to mankind, she glitters from afar,—
> Now the bright evening, now the morning star."

Beauty of Venus.—This planet has an expressive name, and it naturally leads us to expect that the object to which it is applied is a beautiful one. The observer will not be disappointed in this anticipation : he will find Venus the most attractive planet of our system. No such difficulties are encountered in finding Venus as in detecting Mercury ; for the former recedes to a distance of 47° from the Sun, and sometimes remains visible 4½ hours after sunset, as in February 1889. But Venus owes her beauty not so much to favourable position as to surpassing lustre. None of the other planets can compare with her in respect to brilliancy. The giant planet Jupiter is pale beside her, and offers no parallel. Ruddy Mars looks faint in her presence, and does not assume to rivalry.

This planet alternately adorns the morning and evening sky, as she reaches her W. and E. elongations from the Sun. The ancients styled her *Lucifer* (" the harbinger of day ") when a morning star and *Hesperus* when an evening star.

Brilliancy.—Her brightness is such as to lead her to occasionally become a conspicuous object to the naked eye in daytime, and at night she casts a perceptible shadow. This is specially the case near the epoch of her maximum brilliancy, which is attained when the planet is in a crescent form, with an apparent diameter of about 40″, and situated some 5 weeks

L

from inferior conjunction. Though only a fourth part of the disk is then illuminated, it emits more lustre than a greater phase, because the latter occurs at a wider distance from the Earth and when the diameter is much less. Her appearance is sometimes so striking that it is not to be wondered at that people, not well informed as to celestial events, have attributed it to causes of unusual nature. When the planet was visible as a morning star in the autumn of 1887, an idea became prevalent in the popular mind that the " Star of Bethlehem " had returned, and there were many persons who submitted to the inconvenience of rising before daylight to gaze upon a spectacle of such phenomenal import. And they were not disappointed in the expectancy of beholding a star of extreme beauty, though altogether wrong in surrounding it with a halo of mystery and wonder.

At intervals of eight years the elongations of Venus are repeated on nearly the same dates as before, and the planet is presented under very similar conditions. This is because five synodical periods (nearly $=13$ sidereal periods) of Venus are equal to eight terrestrial years. Thus very favourable E. elongations occurred on May 9, 1860, May 7, 1868, May 5, 1876, and May 2, 1884 ; and on April 30, 1892, there will be a similar elongation.

Period &c.—Venus moves round the Sun in an orbit of slight eccentricity, and completes a revolution in 224^{d} 16^{h} 49^{m} 8^{s}. Her mean distance from that luminary is 67,000,000 miles. The apparent diameter of the planet varies from $9'''\cdot5$ at superior to $65''$ at inferior conjunction, and it averages $25''$ at elongations. Her real diameter is 7500 miles. The polar compression is very slight—in fact, not sufficiently decided for measurement ; this is also true of Mercury.

Venus as a Telescopic Object.—When the telescope is directed to Venus it must be admitted that the result hardly justifies the anticipation. Observers are led to believe, from the beauty of her aspect as viewed with the unaided eye, that instrumental power will greatly enhance the picture and reveal more striking appearances than are displayed on less conspicuous planets. But the hope is illusive. The lustre of

Venus is so strong at night that her disk is rarely defined with satisfactory clearness ; there is generally a large amount of glare surrounding it, and our instruments undergo a severe ordeal when their capacities are tested upon this planet. Observations should be undertaken in the daytime, or near the times of sunrise or sunset, when the refulgence of this object does not exert itself in extreme degree. But putting aside the question of definition for the moment, there are other circumstances which conspire to render the view a somewhat unattractive one. There are no dark spots, of bold outline, such as we may plainly discern on Mars, visible on her surface. There is no wonderful arrangement of luminous rings, such as encircle Saturn. There are no signs of dark variegated belts, similar to those which gird both Jupiter and Saturn ; nor is there any system of attendant satellites, such as accompany each of the superior planets. But though Venus is wanting in these respects, she may yet boast an attraction which the outer planets can never display to us, namely, the beautiful crescented phase, which, tradition says, was predicted by Copernicus, and, when afterwards observed in Galilei's telescope, justly considered a convincing fact in support of the Copernican system. The phases are best seen in strong twilight, whenever Venus is favourably situated. It has been asserted that the crescent of this planet has been distinguished with the naked eye ; but the statement is undoubtedly erroneous. Any small glass will show it, however, as it is sometimes well visible when subtending an angle of 50" or 55".

Surface-markings.—In 1666 and the following year J. D. Cassini observed several bright spots on Venus and also two obscure markings ; but the latter were extremely faint and of irregular extent, so that little could be gleaned from them. He watched these forms closely and remarked certain changes in their positions, which finally enabled him to determine the period of the planet's rotation. In 1726 and 1727 Bianchini, at Rome, repeatedly observed dark spots, and their outlines seem to have been so consistent that he depicted them on a chart and gave them names. But J. Cassini, at Paris, failed to confirm these results, though he used telescopes of 82- and

114-feet focus ; and it was supposed the climate of Paris was not suitable for such delicate observations. Schröter reviewed this planet in 1788 and later years, and succeeded in detecting various markings and irregularities in the terminator and cusps. He announced that he had seen the S. horn of the crescent truncated, so that a bright point was apparently isolated at its extremity. From this he concluded there must be mountains of great altitude on the planet, and the perpendicular height of one of these he computed at 22 miles, which is four times the height of the most lofty mountain on the Earth. If the surface of Venus were uniformly level, then her cusps would taper gradually away to points, and no such deformation as that described by Schröter could possibly be produced. And there is strong negative evidence among modern observations as to the existence of abnormal features; so that the presence of very elevated mountains must be regarded as extremely doubtful, if, indeed, the theory has not to be entirely abandoned. The detached point at the S. horn shown in Schröter's telescope was probably a false appearance due to atmospheric disturbances or instrumental defects. Whenever the seeing is indifferent, this planet assumes some treacherous features which are very apt to deceive the observer, especially if his telescope is faulty. Spurious details are seen, which quite disappear from the sharp images obtained in steadier air with a good glass. I have never observed truncation in either of the horns of Venus ; but on certain occasions, when the planet has been ill-defined in passing vapours, it was most easy to believe that a fragment became detached from the extremity of the cusp, just in the manner described by Schröter. But close attention has showed the effect to be false, and revealed its cause. It was the rippling of the image that gave rise to the apparently dissevered cusp, in the same way that passing air-waves and resulting quivers in the image of Saturn's ring will sometimes produce displacements, so that the observer momentarily sees several black divisions, and the edges are multiplied and superimposed one on another. Refraction, exercised by heated vapours in crossing objects, is obviously the source of all this.

Sir W. Herschel frequently examined this planet between 1777 and 1793, but could not discern spots sufficiently definite and durable to enable him to fix the time of rotation. He dissented from Schröter as to the alleged mountains, and said, "No eye which is not considerably better than mine, or assisted by much better telescopes, will ever get a sight of them."

Mädler effected some observations of this planet in 1833 and some subsequent years. He detected spots on two occasions only, but noticed irregularities in the terminator and cusps. Di Vico and others at Rome, in 1840–1, devoted much attention to this object, and secured a large number of observations. They appear to have recovered the spots charted by Bianchini, and described them as of the last degree of faintness. The observers who saw the spots most readily were those who had the most difficulty in detecting the faint companion of a close double star. In the spring of 1841 Di Vico saw a marking on the northern cusp involved in an oval luminosity, and he likened it to a crater on the Moon viewed obliquely. This spot had a diameter of at least $4\frac{1}{2}''$, and it was seen to advance even into the obscure part of the disk.

Rotation-Period.—The following are the periods of rotation as given by the different authorities whose observations we have mentioned :—

1666–7.	Cassini	23 hrs. 21 min.
1726–8.	Bianchini . . .	24 days 8 hrs.[*]
1811.	Schröter . . .	23 hrs. 21 min. 8 sec.[†]
1840–1.	Di Vico . . .	23 hrs. 21 min. 22 sec.

Schiaparelli has recently discussed a large number of observations of this planet, and concludes that, like Mercury, she rotates on her axis in the same time that she completes a

---

[*] This period was probably derived erroneously by Bianchini. It includes 25 periods of 23^{h} 22^{m}, which corresponds with the times of rotation by Cassini and others given in the table.

[†] Schröter's final result. In 1788 he had derived a period of 23^{h} 28^{m} from observations of faint dark spots, and in 1789–91 irregularities in the S. horn of Venus gave him a period of 23^{h} 20^{m} 59^{s}.

sidereal revolution round the Sun, viz. in 224·7 days! I
merely mention this remarkable deduction, without quoting
any facts in opposition to it.

From observations by Perrotin at Nice in 1890, including
74 observations, the rotation of this planet is very slow, and
is made in such a way that the relative positions of the spots
and terminator do not experience any notable change during
many days.

Faintness of the Markings.—Several observers have noticed

1881, Mar. 22, 6ʰ.   1881, Mar. 26, 7ʰ.   1881, Mar. 28, 6¼ʰ.

Venus as an evening star.   (10-inch Reflector; power 212.)

a slight blunting of the S. horn of Venus, and in recent years
dusky spots have not unfrequently been seen, notably by
Buffham, Langdon, and others.   The only markings distinguishable with my 10-inch reflector are faint grey areas,
without definite boundaries.   These are sometimes so delicate
that it is difficult to assign exact form and position to them,
and occasionally I have regarded their very existence as
of doubtful character.   They appear to be mere inequalities in

brightness of the surface, and may be due to different reflective power in parts of the dense atmosphere of this planet. Certainly the spots are nothing like those seen on the disks of Mars and Jupiter, many of which are extremely distinct and show sharply terminated outlines. Dawes, an observer endowed with very keen sight, could never succeed in finding any markings on Venus, and many others have failed. But the evidence affirming their reality is too weighty and too numerously attested to allow them to be set aside. Occasionally the disk appears speckled with minute shadings, and some observers have noticed crateriform objects near the terminator ; but these are uncertain. Brilliant spots have also been recorded quite recently at the cusps.

Perhaps it may be advisable here to add a word of caution to observers not to be hastily drawn to believe the spots are visible in very small glasses. Accounts are sometimes published of very dark and definite markings seen with only 2 or 3 inches aperture. Such assertions are usually unreliable. Could the authors of such statements survey the planet through a good 10- or 12-inch telescope, they would see at once they had been deceived. Some years ago I made a number of observations of Venus with 2-, 3-, and $4\frac{1}{4}$-inch refractors and 4- and 10-inch reflectors, and could readily detect with the small instruments what certainly appeared to be spots of a pronounced nature, but on appealing to the 10-inch reflector, in which the view became immensely improved, the spots quite disappeared, and there remained scarcely more than a suspicion of the faint condensations which usually constitute the only visible markings on the surface. I believe, also, the serrated terminator is not a real feature of the object, but rather an effect either of the rippling contour of the image or of an imperfect or inadequate telescope.

An atmosphere of considerable density probably surrounds this planet, for at the limb the brightness of the disk is much intensified. A medium like this, that reflects and refracts light in extreme degree, is brighter under oblique vision, as at the limb of Venus.

Twilight on Venus.—When Venus is a slender crescent, near inferior conjunction, a feeble luminosity pervades the dark part of the disk similar to the "ashy light" or earth-shine observed on the crescented Moon. On such occasions the unilluminated surface appears to be involved in a phosphorescence. Several observers have, however, described the unilluminated limb of Venus as darker than the background of sky. Zenger, at Prague, has noticed a brownish-red ring surrounding the planet, and he attributes the appearance to much the same cause as that which occasions the coppery colour of the Moon in a total eclipse.

Alleged Satellite.—Cassini, Short, Montaigne, and others, in the 17th and 18th centuries, observed small crescents near Venus and inferred the existence of a satellite; but no such object has presented itself in more recent times. It is extremely probable that the observers were mistaken. In some cases the duplicate image may have been formed by reflection in the eyepiece; in others a small star or planet situated near Venus gave rise to the deception. M. Stroobant has fully investigated this astronomical myth, and disposed of many of the observations, without having recourse to the apocryphal satellite named "Neith" by M. Niesten, who has discussed the question from an affirmative point of view.

Further Observations required.—From the foregoing summary amateurs will notice that several difficult and more or less evanescent features on this brilliant member of our system stand in need of confirmation. Certain disputed forms require also to be looked for. The faint dusky patches, the bright spots at the horns, and the inequalities in the curve of the terminator will sure to be re-observed in future years; and it is necessary that such details should be precisely noted in regard to their positions and outlines as often as possible. A series of reliable observations of this character might enable a fresh value of the rotation-period to be deduced from them; and this is desirable, for though Cassini, Schröter, and Di Vico give periods which are in close harmony, there are elements of uncertainty attached to their results. A new determination of the period would be valuable, and especially so if based on really trustworthy data obtained by one of the best modern telescopes. With the planet situated near inferior conjunction, the crescent (reduced at such a time to a mere thread of light) should be brought into the field, and the observer should look for the extension of a faint glow over the interior parts of the surface, and make comparisons between the relative brightness of the planet's dark limb and of the sky on which it is projected. The telescopic images of Venus are often excellent in daylight, and those who possess means of readily finding the planet at such times will be very likely to gain some useful materials. As to the presumed satellite, that may be relegated to the care of observers who have the leisure and inclination to pursue an *ignis fatuus* ; but should any doubtful object appear in the field with Venus at any time, it ought to be fully recorded and identified, if possible.

Transits of Venus.—Those who were prevented by circumstances of weather or otherwise from witnessing either of the transits of Venus which occurred in 1874 and 1882 lost a spectacle of great rarity, and one which they can never have another chance to behold. The next transit occurs in the year 2004, and its phenomena will doubtless be watched with avidity by the astronomers of a future generation. The transit of 1882 was seen by many observers in England, though in some parts of the country the Sun was obscured by clouds. The planet was distinctly visible to the naked eye as a black circular spot in gradual motion across the solar disk. The most important result of the telescopic observations was of course the re-determination of the Sun's distance ; but amongst the physical features noted, one of the most interesting was the appearance of a silver arc of light on that portion of the planet's edge which was outside the Sun. This is assumed to have been caused by the refraction of an atmosphere on Venus. The phenomenon was seen by several observers, including Prof. Langley in America and Messrs. Prince and Brodie in England.

Occultations of Venus.—An occultation of this planet by the Moon appears to have been recorded by the Chinese on March 19, 361 A.D. Tycho Brahe witnessed a similar phenomenon on May 23, 1587. Mœstlin observed Venus occult Regulus on Sept. 16, 1574 ; and on Oct. 2, 1590, this planet appears to have passed over Mars. Visible occultations of Venus are somewhat rare ; they usually occur in daylight.

A phenomenon of this kind was witnessed on Dec. 8, 1877, over all the W. part of the United States ; and Prof. Pritchett, of Missouri, says :—" The interest taken in it was shared alike by the educated and the illiterate, and even by children." The evening was cloudless, and many persons noted the time of disappearance of Venus as seen by the unassisted eye. With a 12½-inch refractor, power 275, Prof. Pritchett noted that " when the bright limb of Venus was within 8″ or 10″ of the Moon's dark limb, a border of wavering light, several seconds in width, seemed to precede the planet. Its general effect was such as to place in doubt the moment of external contact." A full description of this event, and of the partial occultation of Venus on Oct. 12, 1879, is given in No. 1 of the ' Publications' of the Morrison Observatory, Missouri, U.S.A.

Venus is said, by the Arabian astronomer Ibn-Jounis, to have occulted Regulus on Sept. 9, 885 A.D.; and Hind has examined the observations, by means of Le Verrier's tables of the Sun and planets. He finds that on Sept. 9 in the year mentioned, at 16ʰ 43ᵐ mean time, Venus approached the star within 1′·7 ; so that to the naked eye the latter would appear to be occulted, being overpowered in the glare of the planet.

CHAPTER IX.

MARS.

Appearance of the planet.—Period &c.—Phase.—Surface Configuration.
—Charts and Nomenclature of Mars.—Discovery of two Satellites and of
Canal-shaped markings.—Summary of Observations.—Rotation of Mars.
—Further Observations required.—Changes on Mars.—The two Satellites.
—Occultations of Mars.

Appearance of the Planet.—Mars is the fourth planet in the
order of distance from the Sun. He revolves in an orbit
outside that of the Earth, and is the smallest of the superior
planets. His brilliancy is sometimes considerable when he
occupies a position near to the Earth, and he emits an intense
red light*, which renders his appearance all the more striking.
Ordinarily his lustre does not equal that of Jupiter, though
when favourably placed he becomes a worthy rival of that
orb. In 1719 he shone so brightly and with such a fiery
aspect as to cause a panic. The superstitious notions and
belief in astrological influences prevailing at that time no
doubt gave rise to the popular apprehension that the ruddy
star was an omen of disaster, and thus it was regarded with
feelings of terror. Fortunately the light of science has long
since removed such ideas from amongst us, and celestial
objects, in all their various forms, are contemplated without
misgiving. They are rather welcomed as affording the means
of advancing our knowledge of God's wonderful works as
displayed in the heavens.

Period &c.—Mars revolves round the Sun in 686ᵈ 23ʰ 30ᵐ 41ˢ,
and his mean distance from that luminary is 141,500,000 miles.
The orbit is one of considerable eccentricity, the distance

---

* This was believed by Sir J. Herschel to be due to " an ochrey tinge
in the general soil, like what the Red-Sandstone districts on the Earth
may possibly offer to the inhabitants of Mars, only more decided."

varying between 154,700,000 and 128,360,000 miles. The apparent diameter of the planet when in conjunction with the Sun is only 4″; but this may augment to 30″·4 at an opposition, when the Earth and Mars occupy the least distant parts of their orbits. The real diameter of Mars is nearly 5000 miles.

Phase.—At opposition the disk of Mars is round, but when in quadrature he appears distinctly gibbous and resembles the Moon three days from full. The phase is so palpable that Galilei glimpsed it at the end of 1610. In delineations of Mars the disk is generally drawn circular, the compression being very slight and the phase too trivial to be regarded.

Surface Configuration.—This planet being singularly variable in his position relatively to the Earth, presents at times a diameter so small that the most powerful instruments are ineffective to deal with him. But at certain epochs he becomes an excellent object, with a much expanded disk, on which are displayed a number of bright and dark markings. This happens, however, with comparative rarity ; for only during two months or so near every opposition, occurring at intervals of 780 days, can the planet be well seen. Generally the apparent size of Mars is very inconsiderable, and the disk not sharply defined, especially when the altitude is low. Reliable observations are seldom made at a time far removed from the date of opposition. When the planet was badly placed, in July 1882, an observer secured some observations of position, and published them, thinking he had seen Wells's Comet, which happened to be in the same quarter of the sky !

Mars, in nearer degree than any other member of our system, shows a configuration which may be likened to that of the Earth as regards its permanency ; and in some of its outlines a general resemblance also exists, though in detail there is evidently much that is dissimilar. It is fortunate that the atmosphere of Mars is so rarefied that observers can look upon his real surface-lineaments with satisfactory perspicuity. For more than 250 years now, the telescope has been engaged in perfecting our knowledge of Martian features, and these have exhibited no mobility of form or place (apart from that due to rotation or varying inclination of the planet) so far as may be judged from a comparison of drawings. Plenty of differences exist in the latter, it is true, though similar objects are represented ; but the explanation obviously lies in the inaccuracies of amateur artists, and has little if anything to do with physical changes on the planet.

When the spots were discovered in 1636 by Fontana they were, of course, very dimly glimpsed in the incompetent appliances available at that time. Huygens, in 1659, saw them better by means of his long telescopes, but still very imperfectly. Cassini, in 1666, effected a further advance in the same field, and gathered data from which he was able to announce the period of rotation. His value has proved remarkably correct, considering the means he employed to obtain it and the very short interval over which his inquiries were conducted. Huygens had previously, in 1659, witnessed the returns of a certain spot to the same approximate place on the planet, and was led to infer rotation in either 12^h or 24^h. But this was little better than a guess, and not nearly of the same precision as that which marked Cassini's subsequent determination.

Fig. 31.

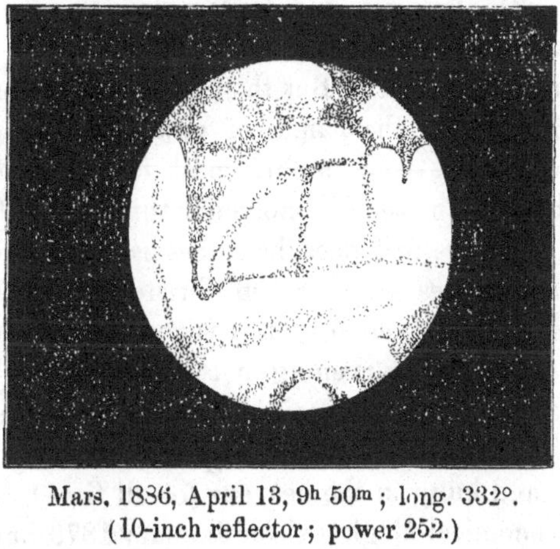

Mars, 1886, April 13, 9^h 50^m ; long. 332°.
(10-inch reflector; power 252.)

Near the poles of Mars are intensely bright patches, which have been considered to be vast areas of snow-crowned surface or fields of ice. These " polar snows " are not situated exactly at the poles, nor are they opposite to each other. Changes affect their aspect. Occasionally these or other bright markings, when on the limb, appear to protrude beyond the disk, and this curious effect of irradiation distorts the limb in a striking manner.

Charts and Nomenclature of Mars.—It is not desirable to trace with any detail the successive labours of those who have chiefly contributed to our knowledge of areographic features. Maraldi, W. Herschel, Schröter, Mädler, Schmidt, and Dawes were foremost amongst the observers of the past; while Schiaparelli and Green are the most successful observers of to-day. As telescopes improved in effectiveness the true forms and characteristics of the markings were discerned, and at the present time some thousands of delineations of this planet must be in existence. Charts of the leading and best-assured features have been formed, and the regions of light and shade (supposed to represent land and sea) have received proper names to distinguish them. Thus there is " Fontana Land," " Maraldi Sea," " Herschel Continent," and others of similar import. Schiaparelli has framed a chart in which the spots are furnished with Latin names taken from classical geography. Mädler's plan was to designate the markings by capital letters of the alphabet, and to divide these by small letters in necessary cases. But the charts of Proctor, Green, and others, in which the names of past and present astronomers are applied, seem to find most favour, though it is admitted that this method of nomenclature is not free from objections. In some instances the names have not been wisely selected. A few years ago, when christening celestial formations was more in fashion than it is now, a man simply had to use a telescope for an evening or two on Mars or the Moon, and spice the relation of his seeings with something in the way of novelty, when his name would be pretty certainly attached to an object and hung in the heavens for all time! A writer in the ' Astronomical Register ' for January 1879 humorously suggested that " the matter should be put into the hands of an advertizing agent " and " made the means of raising a revenue for astronomical purposes." Some men would not object to pay handsomely for the distinction of having their names applied to the seas and continents of Mars or to the craters on the Moon. But it is all very well to disparage a system : can a better one be found ? Probably not; but the lavish use of undeserving names is calculated to bring any system into contempt.

Discovery of Satellites and of Canal-shaped markings.—The interest in this planet has been accentuated in recent years by several circumstances. The discovery of two satellites in 1877 by Prof. Hall, with the 25·8-inch Washington refractor, caused the directors of large instruments to test their capacity upon these minute objects. Schiaparelli's observations of the canal-shaped markings have afforded another attractive feature in

Fig. 32.

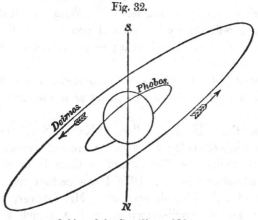

Orbits of the Satellites of Mars.

connection with this planet. He detected a network of dark straight lines stretching generally from N. to S. across the planet, and in the winter of 1881 found these objects duplicated, *i. e.* the lines ran in pairs so close together that they were separated with difficulty. The study of the topography of Mars had never previously revealed structures like these ; yet the Italian astronomer appears to have observed them with " comparative ease whenever the air was still." Other observers have not wholly confirmed the appearances alluded to, but no favourable opposition has occurred since 1877, and no surprise need be felt that the delicate features visible in the

pellucid sky of Italy should elude detection in less genial
climes. In 1886 M. Perrotin, at Nice, using a 15-inch equatoreal, saw a number of the "canals," and some of them were
double. In 1888 the observers having charge of the 36-inch
refractor at Mount Hamilton re-observed the "canals" as
broad bands, but none of them appeared to be duplicated.
The conditions were unfavourable, the planet being more than
three months past opposition.

Prof. Schiaparelli re-observed the duple "canals" in June
1890 with a refractor by Merz of 18 inches aperture, powers
350 and 500. His observations are supported by Mr. A. S.
Williams, of Brighton, who informs me that he detected
forty-three of the "canals," and seven of them were "clearly
and certainly seen to be double." Mr. Williams's instrument
is a 6½-inch reflector by Calver, and powers of 320 and 430
were successfully employed on it ; magnifiers under 300 were
found of little use.

Summary of Observations.—From observations at Bristol I
have drawn up the following summary as to the configuration
of Mars :—

1. That the "Hour-glass" or "Kaiser Sea," and some
other markings of analogous character, present very bold, dark,
and clearly defined outlines, enabling them to be visible in
very small telescopes. In 1873 I saw certain spots with a
refractor of only 1¾-inch aperture. Mr. Grover, in 1867,
"made a set of pencil-drawings, with a 2-inch telescope,
which gave the general markings of the planet very well."
In 'Recreative Science' it is mentioned that on June 7,
1860, a semi-circular dark spot on the N.W. part of the
disk of Mars was distinctly seen with a 1½-inch telescope,
power 120.

2. There is an intricate mass of surface-markings on the
planet, which, in its main features, is capable of being satisfactorily delineated, and which in its general aspect is similar
to the canals depicted by Schiaparelli, though not nearly so
pronounced, straight, and uniform as he has shown in his
charts.

3. The detail is visible in the form of irregular streaks,
condensations, and veins of shading, very faint and delicate

in some parts. The veins apparently connect many of the larger spots, and here and there show condensations, which have sometimes been drawn as isolated spots. A night of good definition, however, reveals the feeble ligaments of shade connecting them.

4. That there exists on the immediate borders of many of the darker patches and veins a remarkable brightness or shimmering, which reminds one of the bright spots merging out of the dark belts on Jupiter. Just contiguous to the " Kaiser Sea," and on its eastern limits, this brightness was so striking in March 1886 as to compare with that exhibited by the N. polar cap. In drawings by many observers these regions of special luminosity have no place, but there is little doubt they occupy a leading position in the physical configuration of Mars.

5. That there is no trace of a dense atmosphere on Mars, as some of the text-books infer. The pronounced aspect of the chief markings, their durableness and continuity of form, the ease with which they may be traced up to the limb, the absence of phenomena indicating dense cloud-bearing air-strata, and other observed facts verify the conclusion that the planet's surface is comparatively free of vapours, and in a totally different condition to that of Jupiter and Saturn.

Rotation of Mars.—The diurnal period of this planet is known with far greater certainty and precision than that of any other planet, the Earth excepted. It will be useful to quote the values derived since Cassini's time :—

		h	m	s
1666.	J. D. Cassini.	24	40	
1704.	J. P. Maraldi	24	39	
1781.	W. Herschel.	24	39	21·7,*
1784.	W. Herschel .	24	37	27

* Herschel's earlier observations were made in 1777–79, and his period, like that of his predecessors, is about 2 min. in excess of the correct value ; but Mädler pointed out that, by giving Mars an additional rotation on his axis, Herschel's value will agree within 2 sec. of his own. Herschel appears to have adopted 768 rotations instead of 769, and may have been

			h	m	s
1838.	J. H. Mädler	.	24	37	23·8
1845.	O. M. Mitchell	.	24	37	20·6
1859.	A. Secchi	.	24	37	35
1864.	F. Kaiser	.	24	37	22·62
1866.	R. Wolf	.	24	37	22·9
1869.	R. A. Proctor	.	24	37	22·735
1873.	F. Kaiser	.	24	37	22·591
1873.	J. F. J. Schmidt	.	24	37	22·57
1883.	A. Marth	.	24	37	22·626
1884.	W. F. Denning	.	24	37	22·34*
1885.	H. G. v. de S. Bakhuyzen	.	24	37	22·66

The last of these, by Prof. Bakhuyzen of Leyden, is probably the best. It was based on a large number of observations extending over 220 years, viz. from those of Huygens in 1659 to those of Schiaparelli in 1879.

In a terrestrial day Mars rotates through 350°·8922, according to Mr. Marth's period. In one hour the axial motion is 14°·6, whereas on Jupiter the horary rate of rotation is 36°·7. At intervals of 40 days (during which Mars completes 39 rotations) the various features on the disk are presented at very nearly the same times as before. Mr. Marth's ephemerides of this planet are extremely useful to those who study the markings ; and these, in combination with the charts and memoirs of Schiaparelli, Green, Terby, and others, greatly facilitate and encourage the renewed study of this object.

Further Observations required.—Favourable oppositions of Mars occur every 15 years, as in 1877 and 1892. It is at such periods that this planet should be sedulously interrogated for new features, or for corroboration of those already known. Rather a high power must be employed—certainly more than 200 ; and if the telescope has an aperture of at least

---

led to this by the excessive periods of Cassini and Maraldi and by the want of intermediate data between his own observations in April 1777 and May–June 1779. His second determination, made in 1784, is more correct.

* Deduced from observations extending over 15 years only, at Bristol.

8 inches, the observer will be sure to discern a considerable extent of detail. He should compare his views with the various charts previously alluded to, and note any inconsistencies. Fresh drawings should also be made ; and if the forms are not well assured on one night, he may confirm them by coming 37 minutes later to his instrument on the following night. Or the collective issue of several nights' work may be included in the same drawing. The bright spots on the planet should be as attentively studied as the darker regions, and given a place in every drawing ; for it is probably in connection with these luminous objects that active changes may be recognized. The " canals " and their duplication form the principal markings to be looked for ; though the successful elucidation of these appearances can only be expected in a case where a powerful telescope, a keen eye, and a good atmosphere operate together. Something of them may be seen under ordinary conditions, and they ought to be very generally sought for by amateurs ; for it is not always that success is found where the best conditions prevail. The great telescopes at Mount Hamilton, Nice, and other observatories may be expected to command some advantages of light, power, and position ; but this need not prevent competition, or induce the idea that common appliances are practically of no avail. Everyone should strive to achieve as much as is consistent with his means and opportunities : indeed there is all the more need for effort and energy in the observer when his tools are seemingly inadequate to a research, and he should endeavour to find, in his own eye and understanding, that power which shall compensate in a great measure for lack of instrumental capacity. Mr. Proctor, in his ' Old and New Astronomy,' has justly remarked :—" The directors of Government observatories have usually been much less successful in studying planetary details than those zealous amateurs who take delight in the study of the heavenly orbs and are ready to wait and watch for favourable opportunities."

Changes on Mars.—Changes have been confidently reported in some of the Martian spots. Instances have been quoted in which particular markings, though very plain at certain times, have scarcely been perceptible at others. Variations in outline as well as in visibility appear to have been witnessed, and the subject is one which merits more extended notice. It has been asserted that the origin of such variations probably lies in the aerial envelope of Mars. In April and May 1888 M. Perrotin, with the great refractor at Nice, failed to re-observe the feature known as the continent "Libya" on Schiaparelli's chart, and stated that though this formation was plainly visible in 1886, it had ceased to exist in 1888. He suggested that the obscuration was really produced by clouds or mists circulating in the atmosphere of Mars. But Prof. Holden reported, from the Lick Observatory, that the object alluded to was distinctly visible with the 36-inch refractor there at the end of July, and in the same form in which it was drawn by Prof. Schiaparelli in 1877–8. It is to be assumed, therefore, that if any change occurred it was one of transient nature.

There are other questions relating to the physical aspect of this planet which future observers should be able to answer. Do the markings retain their distinctness right up to the limb? Is the opaque crescent of the disk (when Mars is in quadrature) involved in any phosphorescence or glow indicating an atmosphere? Are the bright spots and luminous borders to the continents equally as stable as the dark spots, and do they maintain an equable brilliancy?

The N. hemisphere of Mars needs much further study, as it is not so familiarly known as the S. hemisphere. This is due to the circumstance that, at favourable oppositions, the region of the S. pole is suitably presented for observation. It is only when the planet is comparatively distant, and small in diameter, that his N. hemisphere comes into view.

The difference of inclination under which the features are seen at successive oppositions gives rise to many apparent changes of figure. When the S. hemisphere is exposed to the Earth numerous objects are seen which are quite invisible when the opposite hemisphere is displayed to us. These altering conditions have to be considered in their influences by every student of areography.

Satellites of Mars.—After evading the keen and searching

eyes of Sir W. Herschel, and the power of his 40-foot telescope—after eluding the grasp of Lord Rosse's 6-foot speculum, and the frequent scrutiny of Lassell with his 2- and
4-foot mirrors, the two satellites of Mars were ultimately
revealed to Prof. Hall in the 25·8-inch refractor at Washington. These tiny orbs had been enabled to avoid previous
discovery by their minuteness and by their close proximity to
Mars. Yet as soon as they were known to exist many
observers saw them, and in certain cases success was undoubtedly attained with comparatively small instruments.
The late Dr. Erck picked up the outermost satellite with a
$7\frac{1}{3}$-inch objective, and Mr. Pratt saw it with an $8\frac{1}{7}$-inch
mirror by With. But the effect of this eye-straining may
just possibly, in one or two instances, have drawn the imagination out of its normal repose. Mr. Pratt's instrument shows
stars in the group ϵ Lyræ which are invisible in the great
Washington telescope and in the 36-inch mirror formerly
used by Mr. Common; so that it may well have produced a
spectral satellite of Mars. But the satellites are certainly
within the occasional reach of moderate means; for they
were repeatedly seen with a $9\frac{1}{2}$-inch refractor at the Observatory of Princeton, U.S.A., in October and November 1879.
They "were decidedly more easy to see than Mimas," the
innermost satellite of Saturn.

Phobos, the inner satellite, revolves round the planet in
7^h 39^m, in an orbit 6000 miles from the centre of Mars. At
max. elongation the satellite is about 12″ distant from its
primary, and its opposition magnitude is $11\frac{1}{2}$. Deimos, the
outer satellite, revolves in 30^h 18^m, and its orbit is 15,000
miles distant from Mars. Its elongations extend to 32″, and
its opposition mag. is $13\frac{1}{2}$. These diminutive objects are
probably not more than 10 miles in diameter. They are
obviously too faint for common instruments, nor are they
objects on which ordinary amateurs may occupy themselves
with advantage. Of course it forms a highly interesting
spectacle to glimpse, just for once, it may be, the small bodies
which resisted telescopic power for more than two and a half
centuries; but for really useful observations, large aperture
and means of accurate measurement are required.

Occultations of Mars.—The most ancient account of a planetary occultation is probably that given by Aristotle, who refers to a lunar obscuration of Mars that occurred on April 4, 357 B.C., according to the calculations of Kepler. Another occultation of Mars appears to have been recorded by the Chinese on Feb. 14, 69 B.C. Tycho Brahe observed a repetition of the event on Dec. 30, 1595. Mr. Baily describes a phenomenon of this kind which occurred on Feb. 18, 1837, when "the planet appeared of a fine yellow colour both at its ingress and egress. No projection was observed." Mr. Snow, of Ashhurst, saw the occultation of March 12, 1854, and he states the planet "was of almost precisely the same colour as the Moon, and he could not help comparing it to a spangle on the face of the sky. Whilst it was slowly and solemnly vanishing, it gave for several seconds the notion of its being the summit of a lunar mountain, but melted gradually away." As Mars emersed, "nothing whatever was to be seen of the two bodies, clinging together, as it were, by threads of light; nothing of the pear-shaped appearance often recorded as put on by planets under similar circumstances." Mr. J. Tebbutt, of Windsor, N. S. W., watched an occultation of Mars in full daylight on Aug. 12, 1875, when "the rapid disappearance of the planet's disk was an exceedingly interesting phenomenon, its extinction taking place at a considerable distance from the Moon's illuminated disk. The line marked by the Moon's dark limb across the disk was well defined." At the reappearance clouds were prevalent, and "the planet was observed as a small projection on the bright limb;" but he found it difficult to fix the exact time of last contact, owing to the ill-defined character of the planet's gibbous limb. An occultation of Mars was also seen by Prof. Grant at Glasgow on June 3, 1878.

CHAPTER X.

THE PLANETOIDS.

Number.—History of their Discovery.—Dimensions and Brightness.—
Occultation of Vesta.

Number.—These bodies, also called minor planets, and, formerly, asteroids, comprise a very numerous class, and they are
extremely small, being quite invisible to the naked eye except
in one or two special cases. They all revolve in orbits situated
between Mars and Jupiter. The total number discovered is
about 300, of which Prof. J. Palisa of Vienna has found more
than 70, and the late Dr. C. H. F. Peters of Clinton, N.Y., 49.
I have not given exact numbers in the two former cases,
because these discoveries are still rapidly progressing.

History of their Discovery.—The first known planetoid
(Ceres) was sighted by Piazzi on Jan. 1, 1801. The
following year, on March 28, Olbers found another (Pallas).
In 1804, on Sept. 1, Harding discovered a third (Juno);
and in 1807, March 29, Olbers was a second time successful
(Vesta). Then for thirty-eight years no additions were made
to the number. The host of planetoids circulating between
Mars and Jupiter preserved their incognito without disturbance from the prying and wakeful eyes of astronomers.

But in 1845 Hencke, of Driessen, after years of watching,
at length broke the spell of tranquillity by finding another
small planet; and his example was emulated by many other
observers in subsequent years. Hind, De Gasparis, and Goldschmidt were amongst the earliest and most successful of those
who gathered new planets from amongst the stars of the
zodiacal constellations. In later years Luther, Watson, and
Borrelly further extended the list; but Palisa and Peters have
distanced all competitors, and shown a zeal in the work which

has yielded an astonishing aggregate of discoveries. Charlois, at Nice, has latterly earned distinction in the same field.

Since 1845 new planetoids have been found at the rate of more than six per annum, and a rich harvest yet remains to be gathered by the planet-seekers of the future. A very large proportion of those already detected are between the tenth and twelfth magnitudes, and are therefore only to be discerned in good instruments. They present no distinction from small star-like points, and are to be identified by their motions alone. The mythological dictionary has furnished names for them, and they are numbered in the order of their discovery as well.

Dimensions and Brightness.—Vesta is the largest and brightest of the group, while Ceres and Pallas rank as second and third in the same respect. Vesta is about 214 miles in diameter; but the more insignificant members of this family are probably not more than about 15 or 20 miles in diameter. Pallas has the most inclined orbit of all, the inclination amounting to 30° 44'; so that its position is by no means confined to the planet-zone of the ecliptic. Vesta is sometimes brighter than a 6th mag. star; while Ceres, Pallas, and Juno vary between about the 7th and 8th magnitudes, according to their distances from the Earth. A real variation of light has been assumed to occur, but this is not fully proved.

In March 1887 Mr. Backhouse, of Sunderland, saw an apparently new, yellowish-white star near 103 Piscium, and it was just visible to the naked eye. This proved to be Vesta, though the identity of the object was not known at first, and it formed the subject of two Dun Echt circulars.

Formerly, hazy indefinite outlines were attributed to some of the planetoids; but the appearance probably arose from instrumental defects.

The search for these bodies is not a work likely to engage amateurs. Professional observers are best able to grapple with the difficulties attending this kind of observation, where large telescopes, means of exact measurement, and ample data, such as star-charts and ephemerides of the planetoids previously discovered, are requisite. The 'Nautical Almanack' annually contains ephemerides of Ceres, Pallas, Juno, and Vesta ; and observers wishing to pick up any one of them may readily ascertain positions by reference to this work.

Occultation of Vesta.—An occultation of Vesta occurred on Dec. 30, 1871, and it was observed by Mr. C. G. Talmage at Leyton with a 10-inch refractor, power 80. He says the planet was exceedingly bright right up to the Moon's limb.

CHAPTER XI.

JUPITER.

> "Beyond the sphere of Mars, in distant skies,
>   Revolves the mighty magnitude of Jove,
>   With kingly state, the rival of the Sun."

OF all the planets, Jupiter is the most interesting for study by the amateur. It is true that Saturn forms an exquisite object, and that his wonderful ring-system is well calculated to incite admiration as a feature unique in the solar system. But when the two planets come to be repeatedly observed, and the charm of first impressions has worn away, the observer must admit that Jupiter, with his broad disk and constantly changing markings, affords the materials for prolonged study and sustained interest. With Saturn the case is different. His features are apparently quiescent; usually there are no definite spots upon the belts or rings. There is a *sameness* in the telescopic views; and this ultimately leads to a feeling of monotony, which causes the object to be neglected in favour of another where active changes are in visible progress.

Brightness and Position.—Jupiter is a brilliant object in the heavens, his lustre exceeding that of Mars or Saturn, though not equal to that of Venus. I have occasionally seen the planet with the naked eye in the daytime, about half an hour after sunrise; and it has been frequently observed by Bond, in America, with the Sun at a considerable altitude. Humboldt and Bonpland, at Cumana, 10° N. lat., saw Jupiter distinctly with the naked eye, 18 minutes after the Sun had appeared in the horizon, on Sept. 26, 1830. The planet is favourably visible for a considerable time every year, and is only beyond reach near the times of his conjunctions with the Sun, when he usually evades observation for about three months. As regards his altitude, Jupiter becomes exceptionally well placed at intervals of 12 years; thus in 1859, 1870–1, and 1882 his declination was 22° or 23° N., and his height therefore very great when passing the meridian. In 1894 he will occupy a similarly auspicious region to observers in the N. hemisphere. In 1865, 1877, and 1889 his declination was 23° S., and he was favourably presented to southern astronomers.

The image of Jupiter as seen in a telescope is involved in a slight yellow tinge, and with the naked eye the same colour is often apparent. But when observed through a very pure transparent atmosphere, his light nearly approaches the silvery lustre of Venus or the Moon. The planet shines with unusual splendour, considering his great distance from the Sun, and his atmosphere must be highly reflective and possibly intensified by inherent light from the planet himself. The central parts of Jupiter's disk are usually the brightest, as there is a faint shading-off and indefiniteness at the limbs. These and other facts support the view that Jupiter is still incalescent and sufficiently self-luminous to emit a small amount of light.

Period &c.—This planet revolves round the Sun in $4332^{\mathrm{d}}\ 14^{\mathrm{h}}\ 2^{\mathrm{m}}$, which is equal to more than $11\frac{3}{4}$ years. His orbit is somewhat eccentric, so that his distance from the Sun varies from 506,500,000 to 460,000,000 miles, and the mean is 483,300,000 miles. His apparent diameter ranges from a max. of 50″ at a good opposition to 30‴·4 in conjunction. The planet's diameter measured along the equator is 88,000 miles, and the polar compression is very marked, amounting to $\frac{1}{16}$, or, more exactly, to $\frac{1}{15\cdot82}$, according to Engelmann, from a mean derived from eleven observers. When Jupiter is in quadrature there is a slight phase evident in the shading-off of the limb furthest from the Sun.

Belts and Spots on the Planet.—From the time that the telescope became available as a means of astronomical research, it may be readily surmised that an object coming so well within the reach of ordinary appliances, and one displaying so many prominent and variable features, should absorb a large share of attention, and that many facts of interest should have been gleaned as to his physical peculiarities. But it must be confessed that, though something has been learned as to the visible behaviour of the markings, there is much that is perplexing in their curious vagaries. No doubt the vast changes affecting the Jovian envelope, the diversity of the markings, and their proper motions result from the operations of a peculiarly variable atmosphere, affected probably by a heated and active globe beneath it, and by the very rapid movement of rotation to which it is subject.

The telescope, on being turned towards Jupiter, reveals at once an array of dark and light stripes or belts stretching across the disk in a direction parallel to one another and to the equator of the planet. These belts are supposed to have been first detected by Zucchi in 1630. Usually there are two broad and prominent dark belts, one on either side of the equator ; while towards the poles other belts appear, some of them very narrow, partly by the effects of foreshortening. The equatoreal zone of the planet is of a lighter tint, and variegated with white and dark spots and streaks, liable to rapid changes, and indicating that this region is in a highly disturbed condition.

Observations of Hooke, Cassini, and others.—Hooke and Cassini were amongst the first to find definite spots on the surface of Jupiter. From 1664 to 1667 a particularly large and distinct spot was frequently seen in the planet's S. hemisphere. This object disappeared in the latter year, but returned in 1672, and was seen until the close of 1674, when it again temporarily vanished, to reappear at subsequent epochs. Cassini was enabled to determine the rotation-period from this spot. He found that the markings in the immediate vicinity of the equator moved with greater

celerity than those in higher latitudes, the difference in their
rotation-periods being nearly 6 minutes. A century later
Sir W. Herschel confirmed these results : he saw a bright
spot which completed a rotation in nearly 5 minutes less
time than several dark spots. Schröter also made many
observations, and noted frequent changes in the spots and
differences in their rotation-periods. He watched a bright
object near the equator which had a period more than
5 minutes less than some dark spots. In later years Mädler
and others followed up the investigation of these markings,
and with nearly similar results. The various spots were
undoubtedly affected by proper motions, enabling them
to yield discordant rotation-periods. Bright forms near the
equator moved with great rapidity and effected a rotation in
about 9^h 50^m, while dark spots on either side of it occupied
between 9^h 55^m and 9^h 56^m. The markings were evidently controlled by currents of different velocities in the
planet's atmosphere.

Dawes, in 1849 and following years, noted luminous spots,
like satellites in transit, on a belt in the planet's S. hemisphere. In October 1857 he observed a group of eleven of
these objects ; and in 1858 Lassell saw many similar appearances in a bright belt near the equator.

The Ellipse of 1869-70.—In 1869 and 1870 Gledhill, of
Halifax, and Prof. Mayer, of the Lehigh University, saw
a remarkable formation just south of the great belt lying on
the S. side of the equator. It was in the form of a perfect
ellipse, ruddy in colour, and very distinct in outline. Its
major axis was parallel with the belts. It was first observed
on Nov. 14, 1869, and had disappeared in July 1870, though
on Dec. 1, 1871, a similar elliptic ring was seen resting on
the S. equatoreal dark band.

The Red Spot.—In July 1878 a large spot, of oval form
and intense red colour, appeared in about the same latitude
as the ellipse seen by Gledhill and Mayer in 1869-70. It
was first announced by Dennett of Southampton, though it
appears to have been seen a few weeks earlier by Prof.
Pritchett, of Missouri, U.S.A. The object alluded to soon
attracted general notice ; and as it continued visible during

the oppositions of 1879, 1880, and 1881 under the same striking aspect, it created a considerable stir among telescopists, and the "great red spot on Jupiter" became familiarly known both in appearance and in title.

No planetary marking in modern times has enlisted half the amount of attention that has been devoted to this object. It has endured amid all the turmoils of the Jovian atmosphere for twelve years, and has preserved an integrity of form and size which prove it to have been singularly capable of withstanding disruption. But its tint has varied greatly ; so that at times the oval outline of the spot could hardly be discerned amongst the contiguous belts. In the winter of 1881 the interior of the ellipse began to lose tone, and in 1882 it faded rapidly, so that the central region of the spot assumed nearly the same light tint as the outlying bright belts. Apparently the spot had either been filled up with luminous cloudy material or had been partially obscured by the interposition of matter situated higher in the Jovian atmosphere. The elliptical contour of the object was still intact, however, though it had quite lost its bold and prominent character. Only the skeleton of its former self remained, and its entire disappearance seemed imminent. But further decadence was fortunately averted by influences unknown to us, and the spot has continued visible to this day, though shorn of the attributes which roused so much enthusiasm amongst observers more than ten years ago.

From measures at Chicago, in the years from 1879 to 1884, Prof. Hough found the mean dimensions of the spot to be :— Length $11''\cdot75$, breadth $3''\cdot71$. These figures represent a real length of 25,900 miles and a diameter of 8200 miles. The latitude of the spot was $6''\cdot97$ S.

This object has served an important end in attracting widespread observation, not only to itself, but to the general phenomena occurring on the surface of Jupiter. Observers, in studying the red spot, were also led to study the bright equatoreal spots and other features so plentifully distributed over the disk. It was most important this should be done ; for since the time of Herschel and Schröter not much progress had been made in elucidating the proper motions of the

spots and finding an accurate rotation-period for the planet.
Dawes, Lassell, and many others had, it is true, secured some
interesting observations and drawings, but not of the special
kind required, and thus no fresh light had been thrown upon
the vagaries in the behaviour of the spots, as described by
the old observers. But a mass of new facts were now to
be realized. Schmidt at Athens, Prof. Hough at Chicago,
A. Stanley Williams at Brighton, and many others, including myself at Bristol, began systematic observations
of Jupiter, with a view to learn something more of the
periods, changes, and general characteristics of the spots
and other features. The results were of an interesting
nature, though too extensive for more than bare mention
here. In 1879 the red spot gave a rotation-period of
9^h 55^m $34^s·2$, but this increased to 9^h 55^m $35^s·6$ in 1880–1
and to 9^h 55^m $38^s·2$ in 1881–2. During the ensuing three
years the period was almost stationary at 9^h 55^m $39^s·1$, but
in 1885–6 it further augmented to 9^h 55^m $41^s·1$, since which
year it has ranged between 9^h 55^m 40^s and 41^s. From
ten years' observations, the mean period of the red spot
is as nearly as possible 9^h 55^m 39^s.

Bright Equatoreal Spots.—The bright spots near the
equator rotated in 9^h 50^m 6^s in 1880; but in subsequent
years the time slightly increased, for in 1882 I found it
9^h 50^m $8^s·8$, and in 1883 9_h 50^m $11^s·4$. The bright spots
therefore perform a rotation in $5\frac{1}{2}$ minutes less time than the
red spot. The former move so much more swiftly than the
latter that they pass it at the rate of 260 miles per hour, and
in $44\frac{1}{2}$ days have effected a complete circuit of Jupiter
relatively to it. Thus a brilliant white spot, if noticed in
the same longitude as the red spot on one night, will, on
subsequent nights, be observed to the W. of it, and, after an
interval of about $44\frac{1}{2}$ days, the same objects will again
occupy coincident longitude.

Dark Spots in N. hemisphere.—In the autumn of 1880
there was a confluent outbreak of dark spots from a belt in
about 25° N. latitude, and these exhibited a rotation-period
of only 9^h 48^m, so that they travelled more rapidly than the
white spots on the equator. Some short dusky belts were

also remarked slightly S. of the latitude of the red spot, and these indicated a period of 9^h 55^m 18^s. It is clear from these various results that the motion of the Jovian markings does not decrease according to their distance from the equator.

Rotation-Period.—Below are given the times of rotation ascertained by some previous observers:—

		h	m	s
1665. J. D. Cassini	. . . . .	9	55	58
1672. „	. . . .	9	55	50
1692. „	. . . .	*9	50	
1708. J. P. Maraldi	. . . .	9	56	48
1713. „	. . . .	9	56	
1773. Sylvabelle	. . . .	9	56	
1779. †W. Herschel	. . . .	9	54	53 ⎫
	to	9	55	40 ⎭
1779. „	. . . .	*9	50	48 ⎫
	to	9	51	45·6 ⎭
1786. ‡J. H. Schröter	. . . .	9	55	33·6
„ „	. . . .	9	55	17·6
„ „	. . . .	*9	50	27
1835. J. H. Mädler	. . . .	9	55	26·5
„ G. B. Airy	. . . . .	9	55	21·3
1836. J. H. Mädler	. . . .	9	55	23·5
1862. J. F. J. Schmidt	. . .	9	55	25·7
1866. „	. . .	9	55	46·3
1873. O. Lohse	. . . .	9	55	19·6
1880. J. F. J. Schmidt	. . .	9	55	34·4
„ „	. . .	*9	50	
1881. W. F. Denning	. . .	9	48	
„ „	. . .	9	55	17·9
1883. „	. . .	*9	50	8·7
1885. G. W. Hough	. . .	9	55	37·4
„ „	. . . .	*9	50	9
1886. A. Marth	. . . .	9	55	40·6

* Bright spots near the equator of Jupiter.

† Herschel's observations embraced few rotations, and the periods he derived differed considerably.

‡ Schröter also alleges he saw spots return to the same part of the disk in 7^h 7^m, 7^h 36^m, and 8^h 1^m!

1887. A. S. Williams:—   h   m   s
           Spots in 12° N. lat. . .   9 55 36·5
               „   4° N. lat. . . *9 50 40·1
               „   8° S. lat. . . *9 50 22·4
               „   30° S. lat. . .   9 55 17·1
1890. †W. F. Denning 9 55 39

The foregoing list is by no means complete, for, owing to the large number of recent determinations, I have thought it advisable to omit some of them.

It should be mentioned here that the above times of rotation are derived from atmospheric features more or less volatile in nature, and that therefore the actual sphere of Jupiter rotates in a period which we have not precisely discovered. No doubt the motion of the real surface is not very different from that of the atmospheric markings above it. There is reason to think that, whatever the character of the planet's crust may be, we have never yet obtained a glimpse of it. A dense veil of impenetrable vapours appears to surround the globe on all sides, and this is subject to violent derangement from the evolution of heated material or gaseous fluids from the surface below. These disturbances seem to be very durable in some instances as to their observed effects. The atmosphere would, in fact, appear to possess a singular capacity for retaining the impressions of its changes. The permanency of certain spots can hardly be due to continued action from those parts of the disk immediately underlying them; for their variable motions soon transport them far from the places at which they were first seen, and prove their existence to be quite independent of their longitude.

Nature of the Red Spot.—There is much in connection with the red spot that remains in mystery. Its dimensions, form, and motion have severally been ascertained within small limits of error, and the alterations in its tint and degree of visibility have been recorded with every care. But we can only conjecture as to the origin, character, and end of this remarkable formation. What agency produced it,

---

* Bright spots near the equator of Jupiter.
† From ten years' observation of the red spot.

Fig. 33.

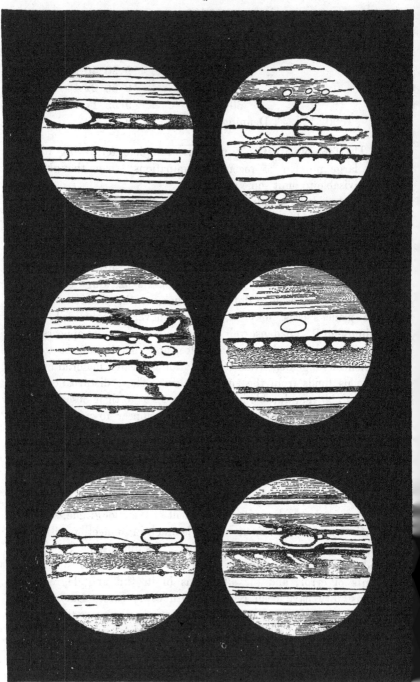

I. 1857, Nov. 27. (Dawes.) II. 1859, Dec. 29. (Huggins.)
III. 1858, Mar. 2. (Huggins.) IV. 1870, Jan. 23. (Gledhill.)
V. 1872, Feb. 2. (Gledhill.)

and moulded the definite elliptical outline it has always preserved—what forces control its oscillations of speed, and keep it suspended so long in the aerial envelope of Jupiter— are matters of pure theory. When, in July 1878, it first came under notice it was a well-developed object, and though Russell in 1876, Lord Rosse and others in 1873, and Gledhill and Mayer in 1869–70 had delineated forms suspiciously like the red spot and situated in the same latitude, yet the several features may not have been absolutely identical, for nothing was seen of the spot in 1877 or in some other years. But there is a strong probability that the red ellipse of 1869–70 must have been the red spot in an incipient stage of its formation. The object may have undergone temporary obscuration, similarly to Cassini's spot two centuries ago.

My own opinion of the spot is that it represents an opening in the atmosphere of Jupiter, through which, in 1878–82, we saw the dense red vapours of his lower strata, if not his actual surface itself. Its lighter tint in recent years is probably due to the filling-in of the cavity by the encroachment of durable clouds in the vicinity. Parts of some of the more prominent belts display an intense red hue like that formerly shown by the red spot, and they may be due to the same causes. Extensive fissures are probably formed in the atmosphere, and quickly distended in longitude by the natural effect of the planet's tremendous velocity of rotation. It is curious, however, that these rents, after a certain distention, assume a durable outline until they lose their colouring and are temporarily if not finally obliterated.

When the red spot was visible under its best conditions I frequently examined it, hoping to detect some mark well in its interior which might serve as a clue to the true rotation-period of the sphere of Jupiter. For if the spot consisted of a clear patch in the planet's atmosphere, I thought it possible some real object on the surface might be discerned through it, in which case the difference in its motion and that of the red spot would enable the rate of motion of the globe to be found. If the spot moves more slowly than the planet, then a surface-marking must appear to pass from the E. to the W. side of the spot ; but no such evidence could be obtained,

owing to the absence of suitable markings. The red tint of
the great spot seemed very general over the entire area of
the ellipse until its central regions paled in 1882. There
were two dark specks, one at the E. and another at the
W. extremity of the spot; but these were unchangeable as
regards position.

The spot, though placed very near the border of the great
S. belt, has never been connected with it, though in Jan.-
Feb. 1884, May 1885, and March–April 1886 the spot

Fig. 34.

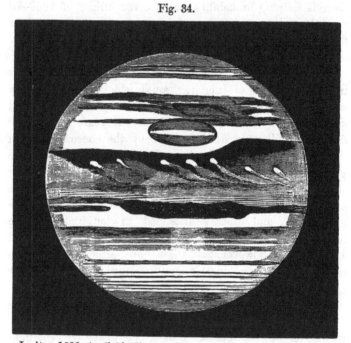

Jupiter, 1886, April 9ᵈ 10ʰ 12ᵐ.   (10-inch reflector; power 252.)

became temporarily attached to a belt on its S. side. There
was some controversy as to this feature, Prof. Hough, from
observations with the 18½-inch refractor at Chicago, alleging
that at no time had the spot coalesced with or been joined to
any belt in its vicinity. But in 1886 many observers succeeded in detecting the junction of the markings alluded to,
and Prof. Young gave a drawing of the appearance as seen

with power 790 on the 23-inch objective at Princeton (see 'Sidereal Messenger,' vol. v. p. 292). The spot and belt were probably at different heights in the Jovian atmosphere, so that there was no commingling of material, one object being simply projected on the other, for the elliptical form of the red spot remained visible all the time. The latter moves more slowly than the connecting belts, and, when clear of them, is often seen with a white aureola fringing its environs.

Bright Equatoreal Spots.—These are affected by rapid changes of form, brightness, and motion. Sometimes they are exceedingly bright; at other periods they are quite invisible. This intermittency is not occasioned (as I assured myself by many observations) by the total extinction of spots and appearance of new ones, but is due to the temporary obscuration of the same objects. The variations are irregular, and probably depend upon phenomena also irregular. The motion of these objects often shows great deviations from their average rate; they are sometimes much in advance of or behind their computed positions. One fine spot of this class was closely watched in 1880 and following years. It was usually in the shape of a brilliant oval, well defined, and occasionally quite as large as the third satellite of Jupiter; but it was sometimes seen as a diffused white patch, apparently emerging from the N. edge of the belt. Whenever the spot was very bright, there was a trail of light or luminous matter running eastwards from it, as though there were an eruption of shining material from the spot, which the rapid rotation of the planet from W. to E. caused to drift in an opposite direction.

Dark Equatoreal Spots.—Closely contiguous to the white spots there are almost invariably seen very dark spots, much deeper in tone than the dark belt upon which they appear to be projected. It has been suggested that these dark spots are shadows from the white spots, which may be elevated formations protruding through the envelope of Jupiter. This idea seems to me untenable; for the dark spots have been distinguished under a vertical Sun, and sometimes they are found one on each side of a white spot. Again, an intensely brilliant spot is occasionally seen without any outlying condensation of dark matter. But though they are not shadows, the dark equatoreal spots certainly have an intimate relation with the brighter markings near them and move with the same velocity.

It is proved from many observations that the longer an object is observed the slower becomes its rate of rotation. Sir W. Herschel found the converse. In discussing his results of 1778 and 1779, he said :—" By a comparison of the different periods it appears that a spot gradually performs its revolutions in less time than at first " (Phil. Trans. 1781, p. 126). But his periods were each based on less than fifty rotations, so that no certain conclusions could be derived from them.

In recent years the rapidly moving bright spots have usually appeared in the equatoreal side of the great S. dark belt. The polar side of the great N. belt also exhibits bright spots, but these rotate in a period only a few seconds less than that of the red spot. Bright spots are also observed to the S. of the latter object and on other portions of the disk.

As to the belts, they are usually straight ; but cases are recorded of slant-belts, in which the direction has been very oblique. One of these was noticed in the planet's N. hemisphere in Mar.–April 1860, and another was seen in the S. hemisphere in Jan. 1872. I observed one near the N. polar shading in Dec. 1881.

New Belts.—The formation of the dark belts seems to be brought about gradually, and they appear to be sustained in certain cases by eruptions of dark matter, which gradually spread out into streams. On Oct. 17, 1880, two dark spots, separated by 20° of longitude, broke out on a belt some 25° N. of the equator. Other spots quickly formed on each side of the pair alluded to, and distended themselves along the belt so that by Dec. 30 they covered three fourths of its entire circumference. At the middle of January the spots formed a complete girdle round the planet ; but they became much fainter, and were soon eradicated by combination with the belt on which they had appeared.

Changes on Jupiter.—Prof. Hough, of Chicago, is adverse to the opinion that rapid changes occur on Jupiter, and mentions the stability of the red spot and other markings in support of his views. He believes that the erroneous statements about sudden changes made by both ancient and modern astronomers are largely due to differences in the telescopic images due to atmospheric variations. No doubt such an explanation will suffice to meet some instances, and the swift rotation of the planet may also have been the unsuspected cause of some of the extraordinary changes described ; but there are real variations as well. These are very frequent in the planet's equatoreal zone.

Further Observations required.—Drawings of Jupiter obtained under the highest powers that may be employed with advantage, and with a cautious regard to faithful delineation, will probably throw much light on the phenomena occurring in this planet's atmosphere. And it is most desirable to pursue the various markings year after year with unflagging perseverance ; for it is only by such means that we can hope to unravel the extraordinary problem which their visible behaviour offers for solution. Too much stress cannot possibly be laid on the necessity of observers being as precise as possible in their records. The times when an object comes to the central meridian should be invariably noted ; for this affords a clue to its longitude, and a means of determining its velocity. Its position, N. or S. of the equator, should be either measured or estimated ; and alterations in tone, figure, or tint described, with a view to ascertain its real character.

The climate of England is very ill-adapted to an investigation of this sort, where the most needful point consists in frequency of observation. If the markings on Jupiter could be re-examined every night, and traced through their changes, an explanation of certain phenomena exhibited by them would soon be forthcoming. The interrupted character of previous observations destroys much of their value. Closely consecutive results are necessary to remove doubts as to the identity of the objects observed; so that, in such a research, natural advantages of position are more desirable than instrumental advantages, for the latter are impotent in a cloudy atmosphere.

The red spot must be watched as long as any vestiges of it remain. Its variations of speed may ultimately yield indications of periodicity*; so may its alterations of tint. The belts in the vicinity of the spot demand an equal share of attention ; for it may be possible to divine from their changes whether there are any links of association between them and the red spot. In recent years the latter has apparently repulsed the belts on its N. side, though suffering encroachments from those on its S. side.

The equatoreal spots also deserve continued vigilance on the part of observers. It has already been stated that the bright spots vary rapidly ; their motions are not uniform in rate, and what is now wanted is a large number of new observations. Does accelerated velocity occur with increased brilliancy of these objects? Are their alternating disappearances and revivals uniform in period ? and are they really due to transitory obscurations of the same durable forms? Are the dark spots which frequently border the white spots implicated in effacing the latter? Many other questions like these are suggested by the curious behaviour of the markings, and the discriminating observer will know how to gather the materials likely to aid in answering them. The rotation-period has been already found in regard to many features ; but this element may be re-investigated with profit, for the velocity of the spots offers a very complex problem for solution. Do the markings generally exhibit a retardation of speed as long as they subsist? Abnormal spots, such as those which made their apparition in the autumn of 1880, should be traced through any vagaries they may present ; and peculiar shape or direction in the belts will also merit study, as possibly supplying facts of consequence. It will be important to learn whether objects in a certain latitude have a common rotation-period, or whether different forms give different times. The rate of

---

* The question of periodicity is an extremely interesting one as affecting the disposition, form, and colours of the markings on Jupiter. Certain features visible in 1869-70 were unmistakably reproduced in 1880, and it has been suspected that the cycle of these changes accords with the length of the Jovian year. Future observations must be compared with old drawings and records for the identification of similar features if they are recurrent.

motion shown by certain features may depend upon their character, and not so much upon their position in latitude.

The altitudes of the various markings affords another promising line of research. The appearances and changes of closely contiguous features may be expected to furnish useful data in this connection. Owing to their proper motions they apparently overlap each other at times, and in their alterations of aspect the observer may discover the clue to their relative heights. The subject is discussed in a practical and interesting way by Mr. Green (Memoirs R. A. S. vol. xlix. p. 264) and by Mr. Stanley Williams (' Zenographic Fragments,' i. p. 112), and these works should be consulted by everyone engaged in the study of Jovian phenomena.

It is unfortunate that the observer, in delineating this object, must perforce adopt an extremely hurried method of representing what he sees at the telescope. The planet turns so quickly upon his axis that forms near the central meridian become sensibly displaced in a few minutes ; indeed, it has been stated that an interval of two minutes only is sufficient to introduce a change obvious to simple eye-estimation. In order, therefore, to complete a sketch, the utmost dispatch is requisite ; for this object cannot be depicted from the combined outcome of several evenings of observation. The proper motions of the different features prevent this. With Mars, or any orb exhibiting markings relatively constant, collective results are extremely valuable, and more trustworthy than pictures depending upon an isolated observation.

Amateurs, in entering upon these observations, should be prepared for rapid changes in the apparent aspect of Jupiter caused by his rotation, and not hastily infer them to be real. They should also hesitate before placing confidence in any anomalous results obtained under indifferent seeing ; for bad images have been directly responsible for many misleading announcements.

Occultations of Jupiter by the Moon.—Phenomena of this kind are always awaited with keen interest by the possessors of telescopes ; but it is rarely that all the circumstances are favourable. The first recorded instance appears to have been in A.D. 847. In 1792, on April 7, Schröter observed an occultation of this planet, with a desire to verify his suspicion of a lunar atmosphere. He saw that " some of the satellites became indistinct at the limb of the Moon, while others did not suffer any change of colour. The belts and spots of Jupiter appeared perfectly distinct when close to the limb of the Moon." On Jan. 2, 1857, an occultation took place under conditions which rendered it visible to many observers in this country, and the most interesting fact elicited was that at emersion a dark border was seen attached to the arc of the Moon projected on the planet. Mr. Lassell described this dark border as " a shadowy line, in character, magnitude, and intensity extremely like Saturn's obscure ring projected on the ball." During the thirty years following 1859 only two occultations visible in England occurred, and the last of these, on August 7, 1889, was widely observed. On this occasion Capt. Noble and others re-detected the shadowed edge of the Moon seen by Lassell in 1857. " It was a

Fig. 35.

Occultation of Jupiter, Aug. 7, 1889.

strongly marked shading, following the outline of the Moon's limb." At Bristol I recorded that, at the disappearance, the outer margin of our satellite was fringed with light where it crossed the planet ; but at the reappearance this effect had vanished, and the appearance was perfectly normal. The disk of Jupiter, where it met the edge of the lunar disk, looked dusky by the effects of contrast ; but I saw no marked shading with a sharply terminating boundary, such as appears to have been remarked elsewhere. As the planet emerged definition was superb, the belts were lividly distinct, and the spectacle was one of the prettiest that could be imagined. The red spot was going off the W. limb, and the disk was covered with belts ; many of them near the poles were extremely narrow, like fine lines drawn with a sharp lead pencil. I used a 4-inch refractor, powers 65 and 145, with this instrument the foregoing sketch was made. The exceptional distinctness of the Jovian markings on this occasion shows that the proximity of the Moon has certainly no tendency to efface planetary details, but rather to intensify them*.

On Sept. 3, 1889, an occultation of Jupiter was visible in America, and observed by Mr. Brooks at Geneva, N.Y., with a 10⅜-inch equatoreal. His drawing, made from a photograph and eye-observations, shows nothing of a dark fringe bordering the Moon's limb.

Fig. 36.

Jupiter and satellites seen in a small glass.

The four Satellites.—When Galilei directed his telescope to Jupiter on the evening of Jan. 7, 1610, he saw three small star-like points near the planet ; so :—

On Jan. 13 he discovered a fourth ; thus :—

and ascertaining that these bodies followed Jupiter in his course, concluded them to be moons in attendance upon him.

* On the morning of Dec. 5, 1887, I made a drawing of Saturn, the image of the planet being remarkably well defined, though the Moon was only 1° distant.

At first the discovery was discredited by others ; but it soon had to be accepted as an incontestable fact of observation. These satellites are usually among the very first objects which the amateur views in his telescope, and they form, in combination with their primary, an exquisite picture, the impression of which is not soon forgotten. The periods, distances, &c. of the satellites are as follows :—

No. and Name.	Mean Distance.		Sidereal Period.	Mean Apparent Diameter.	Real Diameter, in miles.
	Diameters of Jupiter.	Miles.			
I. Io	3·03	267,000	h m s 1 18 29	1·02	2390
II. Europa	4·72	425,000	3 13 18	0·91	2120
III. Ganymede	7·71	678,000	7 4 0	1·49	3480
IV. Callisto............	13·55	1,193,000	16 18 5	1·27	2970

The third satellite is much the largest, and its brightness is about equal to that of a star of the 6th mag. The other three may be rated as generally 7th mag., though their brightness is variable, especially that of the fourth satellite, which has been seen exceedingly faint.

It is customary to distinguish these objects, not by their names, as in the case of the moons of Mars, Saturn, and Uranus, but by the Roman numbers affixed to them progressively according to their distances from Jupiter.

The satellites are just visible to the naked eye when the conditions favour their detection ; but they are so much involved in the rays of the planet, and often so near to him, that it may be regarded as an exceptional feat to discern them without telescopic aid. When III. and IV. are near their max. elongation and on the same side of the planet, they have been occasionally observed separately. I. and II., though much closer to Jupiter and more within the influence of his glare and rays, have been similarly seen. When attempting such observations it is best to hide the bare disk of the planet behind some terrestrial object, as this will cut off the obnoxious rays and prevent the brilliant light from dazzling the eye. An opera-glass, or any small portable telescope, reveals the whole retinue of satellites, and enables them to be traced through their revolutions. The ' Nautical Almanack' gives diagrams of their diurnal positions, and with this work as a reference observers will find no difficulty in identifying them apart.

Sir W. Herschel, in the years 1794 to 1796, found that the satellites revolve on their axes in the same time that they revolve about the planet. He was led to this conclusion by a study of the variations in the light emitted by the satellites in different parts of their orbits, and described I. as " of a very intense bright, white, and shining light,—brighter than II. or IV. (not larger). IV. inclines to red, and nearly as bright as II. The latter is of a dull ash-colour. III. is very white." Modern observers have selected II. as relatively the most highly reflective, while IV. is the least. Spots exist on the surfaces of these objects, and probably occasion many of the differences observed.

The eclipses, occultations, and transits of the satellites afford a very fertile and attractive series of phenomena for telescopic review. The exact times of occurrence are tabulated in the ' Nautical Almanack' and asterisks are affixed to such as are visible in this country. Prior to the date of opposition of Jupiter the eclipses occur of course on the W. side of the disk, while after opposition they take place on the E. side. The durations are as follow for the several satellites :—I. = 2^h 20^m, II. = 2^h 56^m, III. = 3^h 43^m, IV. = 4^h 56^m. In reference to III. and IV. the entire phenomenon may be generally observed ; but this is not so in regard to II., as the emersions are frequently effected behind the planet. Only the immersions of I. are visible before opposition, from the same cause; for the satellite enters the cone of shadow close to the planet's limb, and only comes out of it when the globe of the planet is interposed in the line of sight. In such cases the satellite emerges soon after from the limb of Jupiter ; so that its obscuration has been compounded of two separate phenomena, viz. an eclipse and an occultation. After opposition this satellite is first occulted and then eclipsed. IV. sometimes escapes eclipse altogether, by passing above or below the shadow.

The motion of light was discovered, and its velocity determined, by means of the eclipses of Jupiter's satellites. These
phenomena are also useful in ascertaining longitudes. A
spectator on Jupiter himself would see a vast number of solar
and lunar eclipses—about 4500 of each—during the Jovian
year of 4332·6 days, because the three inner satellites exhibit
these phenomena at every revolution, their orbits being very
slightly inclined to Jupiter's equator, and the latter being but
little inclined to the plane of the ecliptic.

The occultations of the satellites are comparatively frequent,
and may be well observed in a good telescope. A tolerably
high magnifier is required to witness these occurrences with
the best effect, the disks of the satellites being small and not
clearly traceable through the various stages of their disappearances unless much amplified. With considerable telescopic power the disks are well seen, and it then becomes
feasible to watch the satellites, first as they come into contact
with the limb, then as the globe of the planet overlaps more
and more of their diminutive forms, and finally as they reach
last contact and withdraw their narrow unobscured segments
behind the expansive sphere of their primary. Both the
beginning and end phase of these occultations is generally
observable in regard to Sat. IV., and frequently also in the
case of III. But with reference to II. and I. it often happens
that only the disappearance or reappearance can be witnessed.
These occultations have furnished some singular and unex-
plained facts of observation. On meeting the limb of Jupiter
Sats. I. and II. have not always disappeared in a normal way.
On April 26, 1863, Wray, with an 8-inch objective, saw II.
distinctly projected within the limb for nearly 20ˢ. Other
similar cases are recorded. The satellites have been seen
apparently " through the edge of the disk." One observer
mentions that II. appeared and disappeared several times
before occultation. The explanation appears to be that there
is so much irradiation round the disk of Jupiter that it pro-
duces a false limb, and it is through this the satellites have
been seen. A very tremulous image, in bad air, may also be
responsible for some of the anomalies recorded.

The transits of the Jovian moons offer the most attractive

phenomena of all, and they come well within the reach of small telescopes. On entering upon the planet they are visible as bright round spots projected on the dusky limb, and subsequently present some eccentric features. II. is invisible, except on the limbs ; I. is often seen as a grey spot threading along the belts ; III. appears as a large dark spot *, nearly as black as its shadow ; IV. seems to be black, and scarcely to be recognized from its shadow. The appearances are certainly to some extent variable. Mr. Stanley Williams has seen III. as a *brilliant* disk at mid-transit. I. sometimes crosses the whole disk as a white spot; at certain other times it is invisible ; at others, again, it is seen as a faint grey spot. IV. is not always black, its aspect depending upon the chord it traverses. Thus, on the evening of Sept. 12, 1889, Mr. Williams, Mr. G. T. Davis of Reading, and myself were observing Jupiter when IV. was in transit on a belt in the N. hemisphere, but not a vestige of the satellite was seen by any of us. On the morning of May 23, 1890, at 3^h 30^m A.M., however, while observing the red spot on Jupiter, I noticed a black circular spot on the great N. equatoreal belt ; and this proved to be IV. in transit. These peculiarities have been accounted for as partly due to contrast and partly to dusky spots on the surfaces of the satellites. Dr. Spitta has made a number of experiments to elucidate this subject, and concludes that " the perpetual whiteness of the second satellite, and the darkened tints of the others during transit, are due to differences in their relative albedo [reflective power] as compared with that of Jupiter, and are not dependent upon the relative quantity of light reflected by one or the other, or upon any physical peculiarities of the Jovian system."

The shadows of the satellites transit the disk as dark spots larger than the satellites themselves, owing to the penumbral fringes. Before opposition these shadows precede the satellites; after opposition the latter come first. The shadow of II. appears to be much lighter than the others, and is usually of

---

* Amongst the first observers of these dark transits were Cassini (Sept. 2, 1665), Romer (1677), and Maraldi (1707).

a pale chocolate-colour ; and I saw it thus at the opening of the year 1885 :—

Fig. 37.

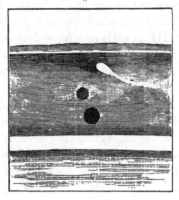

Shadows of Jupiter's Satellites II. and III. near an equatorea white spot (Jan. 1, 1885, 7ʰ 20ᵐ A.M.).

Sat. II. is probably involved in an atmosphere sufficiently dense to enable it to present undue luminosity relatively to the others ; and if so, the feeble shadow it transmits on Jupiter may be partly explained by the effects of refraction. On the day of opposition both satellites and shadows are projected on the same part of the disk, and the latter are occulted by the former.   On Jan. 14, 1872, Mr. F. M. Newton saw I. centrally placed on its shadow ; so that the satellite was apparently surrounded with a ring of shade.   On May 13, 1876, Mr. G. D. Hirst saw Sat. I. partly occulting its own shadow ; a black crescent was seen in the bright zone N. of the equator.   On Feb. 18, 1885, Dr. R. Copeland, at Dun Echt, saw the shadow of I. " almost totally occulted by the satellite itself ; as the satellite approached Jupiter's limb it came out quite bright and large, with a mere crescent of the shadow showing on its southern edge."   This phenomenon was also observed at Bristol.

Occasionally all the satellites become invisible at the same time, being either eclipsed, occulted, or in transit.   An instance of this kind was recorded by Molyneux on Nov. 2, 1681 (O.S.). Sir W. Herschel observed a similar occurrence on May 23,

1802 ; also Wallis on April 15, 1826, and Dawes and others on Sept. 27, 1843, and Aug. 21, 1867. A visible repetition of the event was narrowly avoided on the morning of Oct. 15, 1883. On this occasion the planet should, according to the 'Nautical Almanack,' have been denuded of his satellites for a period of 19 minutes ; but this disappearance did not occur, for at no time were all the satellites included within the margin or shadow of Jupiter. I observed that Sat. III. entered upon the disk just as IV. released itself, and the two formed a curious configuration at 4ʰ A.M., hanging close upon the planet's limbs.

Spots have been seen on the satellites both in transit and while shining on the dark sky. This particularly refers to III. and IV. II. has never given indications of such markings on its bright uniformly clear surface. Dawes, Lassell, and Secchi frequently observed and drew the spots. Secchi described III. as similar in aspect to the mottled disk of Mars as seen in a small telescope ; his drawings exhibit no analogy, however, to those by Dawes of the same object. III. has been remarked of a curious shape, as if dark spots obliterated part of the limbs. Sat. I. was observed in transit on Sept. 8, 1890 by Barnard and Burnham, and it appeared to be *double*, being divided by a bright interval or belt. They used a 12-inch refractor, powers 500 and 700, and the seeing was very fine.

Many other curious points have been noticed in the various aspects and phenomena of the Jovian satellites. Further observations will doubtless throw new light on some of the puzzling records of the past.

Occultation of a Star.—An occultation of the 7th mag. star 4 Geminorum by Jupiter took place on Nov. 7, 1882, and it was observed by Prof. Pritchett, of Glasgow, Missouri, with a 12¼-inch equatoreal, power 200. "The images of both planet and star were very steady. The margin of Jupiter's disk was very sharply defined. The immersion was very near the N. border of the broad S. equatoreal belt. At 11ʰ 28ᵐ 10ˢ·65 local mean time the star was *apparently* within the dark outline of the disk, apparent geometric contact having occurred at 11ʰ 20ᵐ 24ˢ·49. For a moment the star seemed to disappear, but a moment later was plainly seen, as if through a well-defined notch in the otherwise *continuously even* margin. This notch lasted $46^s\cdot26$, and at $11^h\ 28^m\ 56^s\cdot91$ it *vanished*, and the light of the star was *entirely extinguished*." The emersion of the star could not be observed, as clouds supervened.

CHAPTER XII.

SATURN.

Apparent lustre.—Grand spectacle afforded by the Rings.—Period &c.
—"Square-shouldered" aspect.—Early Observations.—Belts and Spots on
the Planet.—Rotation-Period.—The Rings.—Divisions in the outer Ring.
—The transparent or Crape-ring.—Discordant Observations.—Eccentric
position of the Rings.—Aspect.—Further Observations required.—Occultations of Saturn.—The Satellites.—Occultations of Stars by Saturn.

> " Muse, raise thy voice, mysterious truth to sing,
>   How o'er the copious orb a lucid ring,
>   Opaque and broad, is seen its arch to spread
>   Round the big globe, at stated periods led."

THIS planet shines brighter than an ordinary first-magnitude
star, and is a pretty conspicuous object, though less luminous
than either Venus, Jupiter, or Mars. He emits a dull
yellowish light, steadier than the sparkling lustre of Mercury
or Venus.

The globe of Saturn is surrounded by a system of highly
reflective rings, giving to the planet a character of form
which finds no parallel among the other orbs of our system.
His peculiar construction is well calculated to be attractive in
the highest degree to all those who take delight in viewing
the wonders of the heavens. Saturn is justly considered one
of the most charming pictures which the telescope unfolds.
A person who for the first time beholds the planet, encircled
in his rings and surrounded by his moons, can hardly subdue
an exclamation of surprise and wonder at a spectacle as
unique as it is magnificent. Even old observers, who again
and again return to the contemplation of this remarkable
orb, confess they do so unwearyingly, because they find no
parallel elsewhere ; the beautifully curving outline of the
symmetrical image always retains its interest, and refreshes
them with thoughts of the Divine Architect who framed it !

The luminous system of rings attending this planet not only gratifies the eye but gives rise to entertaining speculations as to its origin, character, and purposes with regard to the globe of Saturn. Why, it has been asked, was this planet alone endowed with so novel an appendage? and what particular design does it fulfil in the economy of Saturn? It cannot be regarded as simply an ornament in the firmament, but must subserve important ends, though these may not yet have been revealed to the eye of our understanding.

Period &c.—Saturn revolves round the Sun in 10,759 days 5 hrs. 16 min., which is equal to nearly $29\frac{1}{2}$ years. His mean distance from the Sun is 886,000,000 miles, but this interval varies from 841 to 931 millions, owing to the eccentricity of his orbit. When in opposition his apparent diameter reaches $20''\cdot7$, and declines to $15''$ at the time of conjunction. The planet's actual diameter is 75,000 miles, and his polar compression very considerable, viz. about $\frac{1}{10}$, which exceeds that of any other planet. His synodic period is equal to 378 days; so that he comes into opposition with the Sun thirteen days later every year. The oblate figure of his disk is very noticeable when the rings are turned edgeways to the Earth and practically invisible; but when they are inclined the complete contour of the globe is lost, and the polar flattening becomes scarcely obvious.

" *Square-shouldered* " *Aspect.*—Sir W. Herschel, from observations in April 1805, said :—" There is a singularity which distinguishes the figure of Saturn from that of all the other planets." On April 19 of the year named he described the planet as " like a parallelogram with the four corners rounded off deeply, but not so much as to bring it to a spheroid." This gave the globe a " square-shouldered " aspect. But this curious figure appears to have been very rarely observed in subsequent years; and accurate measures with the micrometer were adduced in 1833–48 in proof that no such anomaly had a real existence. Dr. Kitchiner, commenting on Herschel's remarks, said :—" I have occasionally observed this planet during thirty years, and I do not remember to have seen the body of it of this singular form except for a few months about September 1818." But there is no doubt that occasionally the planet *does* assume an *apparent* form similar to that attributed to it by Herschel. In the autumn of 1880 I studied the visible appearance of Saturn by means of a 10-inch reflector, and recorded as follows :—" The S. pole, over which the dark belts lay, seemed compressed in the most remarkable manner ; but where a bright belt intervened, in about lat. 45°, the contrary effect was produced. Here the limbs were apparently raised (by irradiation) above the spherical contour; so that the distorted image gave the planet that distinctly ' square-shouldered ' aspect sometimes mentioned in text-books." The explanation appears to me very simple. The singular figure is due to the contrasting effects of the belts. While the bright belt in lat. 45° causes a very evident shouldering-out of the limbs at its extremities, the dark belts nearer the pole and the equator act with opposite effect, for they apparently compress the disk where they meet the limbs, and thus the eye discerns a figure to all appearance distorted into the " square-shouldered " form. Mr. J. L. McCance confirmed these remarks by independent observations at the same period with a 10-inch reflector by Calver (' Monthly Notices,' vol. xli. pp. 84, 282).

Early Observations.—The appearance of Saturn offered a considerable difficulty to observers soon after the invention of the telescope. Galilei became greatly perplexed. He saw the planet, not as a circular globe like Jupiter, but distinctly elongated in shape, and conceived the appearance to be due to a central globe with smaller spheres hanging on the sides ! He continued his observations, without, however, arriving at the solution of the mystery, until the malformation began to dis-appear ; and in 1612 he was astonished to find the disk spherical. In his surprise, he asked—" Were the appearances indeed illusion and fraud, with which the glasses have so long deceived me, as well as many others to whom I have shown them ? . . . The shortness of the time, the unexpected nature of the event, the weakness of my understanding, and the fear of being mistaken, have greatly confounded me." Gassendi, in 1633, also announced that Saturn appeared to him to be closely attended by two globes of the same colour as the planet. Riccioli alleged that the planet was surrounded by a thin,

plain, elliptic ring, connected with the sphere by two arms.
None of Galilei's contemporaries possessed the instrumental
means to extricate him from his doubts; and it remained for
Huygens, in 1654 (twelve years after the death of Galilei), to
discover that Saturn "is surrounded by a slender flat ring,
which in no part coheres with the body of the planet, and is
inclined to the ecliptic"*. The same observer showed that
the disappearance which had so puzzled Galilei arose from the
varying inclination in the ring: at times it would become
invisible, when presenting its narrow edge to the Earth, and
this actually occurred again in 1671, as Huygens had predicted. In 1676 Cassini detected a belt upon the planet, and
also a dark division in the ring. Dr. Smith's 'Optics' (1738)
thus alludes to these discoveries :—

"In the year 1676, after Saturn had emerged from the
Sun's rays, Sig. Cassini saw him in the morning twilight

Fig. 38.

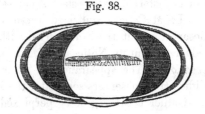

Saturn, as observed by Cassini in August 1676.

with a darkish belt upon his globe, parallel to the long axis
of his ring as usual. But what was most remarkable, the
broad side of the ring was bisected right round by a dark
elliptical line, dividing it, as it were, into two rings, of which
the inner ring appeared brighter than the outer one, with
nearly the like difference in brightness as between that of
silver polished and unpolished—which, though never observed before, was seen many times after with tubes of
34 and 20 feet, and more evidently in the twilight or moonlight than in a darker sky."

From the time when Galilei's inadequate glass revealed the

* Huygens appears to have used a refractor of 2⅓-inch aperture and
23-feet focal length, with a power of 100, in effecting this discovery.

" threefold " aspect of Saturn, and led up to Huygens's solution of the mystery in 1654, this planet has been successively interrogated with the improved telescopes which every generation has produced. Cassini, W. Herschel, Encke, Bond, Lassell, Dawes, and Hall are names familiar to us as having materially advanced our knowledge of this unique orb, both as to his surface-configuration and as to his numerous retinue of satellites.

Belts and Spots on the Planet.—Parallel belts are seen on the surface of Saturn, but they are much fainter than those on Jupiter, and they seldom display the spots and other irregularities interspersed with the belts of the latter planet. Well-bounded spots have rarely been distinguished on the disk of Saturn ; the belts normally appear equal in tone, without breaks, condensations, abrupt curves, or branches, so that the rotation-period has only been accurately determined by Herschel and Hall. And in these cases the markings were certainly atmospheric, and probably affected by proper motions similar to those operating on Jupiter.

Cassini and Fatio remarked two bright streaks on the planet as early as 1683. Sir W. Herschel, in 1790, observed a very dark spot near the margin of the limb, and a few modern observers have been successful in distinguishing either bright or dark spots or patches, though no continuous and useful observations appear to have been secured. In the winter of 1793 Herschel noticed a very distinct quintuple belt, which consisted of three dusky and two intervening light zones. The dark belts presented a dusky yellow hue, while the spaces separating them were white. He recognized the evidences of rotation in the quintuple belt ; for on the same nights, after a few hours' interval, it exhibited considerable variation. Though seen with great precision at first, it became indistinct at a later hour, and the individual belts were placed at unequal distances.

Rotation-Period.—Prof. A. Hall, at Washington, discovered on Dec. 7, 1876, a well-defined white spot, 2″ or 3″ in diameter, and situated just below the ring of Saturn. He watched this object till Jan. 2 following, when it had become faint and indistinct, and the planet being low and the weather unfavourable no further observations were made. The spot had fortunately been seen at four other observatories in the United States, Prof. Hall having notified its existence to them; and on discussing the results, a rotation-period was found not differing largely from Herschel's value derived from the quintuple belt in 1793. These are, in fact, the only two determinations on which we may place confidence. They are as below:—

		h	m	s	Probable error.
1793.	Sir W. Herschel . .	10	16	0·4	2 min.
1877.	Prof. A. Hall . . .	10	14	23·8	2·3 sec.

Schröter, from different spots, computed periods of 11^h 40^m 30^s, 11^h 51^m, and more than 12^h; but these are probably excessive. The difference of 1^m 37^s between the values of Herschel and Hall is relatively a trivial one, as the markings observed were doubtless atmospheric and subject to irregularities of motion. As to the rotation of the ring, Herschel, in 1789, detected some bright marks on it, and deduced the period as 10^h 32^m $15^s·4$[*]. Many astronomical works give the rotation-period of Saturn as 10^h 29^m $16^s·8$; and this is adopted in Chambers's ' Descriptive Astronomy,' 4th edit. vol. i. p. 653. The mistake has its origin in Laplace's *Système du Monde*, where it is stated that Saturn rotates in 0·428 of a day, and the ring in 0·437, which, reduced to hours, minutes, and seconds, give 10^h 16^m $17^s·2$ and 10^h 29^m $16^s·8$.

The equator of Saturn is usually the brightest part of the disk. On its S. side, in recent years, it has been bounded by a very dark narrow belt. Further S. the whole disk seems involved in a faint shading, of a decidedly yellowish hue.

---

[*] Schröter, Harding, Schwabe, and others have observed luminous points on the rings, but they have remained stationary, so that the period of rotation announced by Herschel has never been confirmed, but rather disproved by counter-evidence. Herschel wrote, in November 1789:— " I formerly supposed the surface of the ring to be rough, owing to luminous points like mountains seen on the ring, till one of these supposed luminous points was kind enough to venture off the edge of the ring and appear as a satellite. I have always found these appearances to be due to satellites."

Sometimes a considerable number of belts are visible ; but they are evidently liable to changes, so that the same number and arrangement are not preserved from year to year.

The Rings.—As to the luminous rings, the extreme diameter of the outer one is about 40″, or more than 170,000 miles; and the black division, separating it from the inner one, is 0‴·4, or 1700 miles. The outer ring has a breadth of 2‴·4, or 10,000 miles ; while the inner one measures 3″·9, or 17,000 miles. The outer ring is less luminous than the inner ; the

Fig. 39.

Saturn, 1885, Dec. 23, 7ʰ 54ᵐ. (10-inch reflector, power 252.)

latter, round its outer edges, is extremely brilliant, and has sometimes been described as the brightest part of the Saturnian system. The inner part of this ring is much shaded-off, and offers a strong contrast to the silvery whiteness of the other portion.

Divisions in the Outer Ring.—In the middle of the eighteenth century Short, the optician, using one of his excellent reflectors, thought he saw the outer ring divided by several dark lines; but no other observer confirmed his suspicion. In the third decade of the present century Quetelet and Capt. Kater appear to have observed Short's divisions, but Sir J. Herschel and Struve looked for them in vain. In 1837 Encke fully satisfied himself, by several observations and measurements, as to the objective existence of the divided outer ring. The division was not central, cutting the ring into equal parts, but situated in the inner part of the ring, so that the wider part was outermost. In subsequent years this division has been sometimes seen and placed nearest the outer edge of the ring. Certain observers, provided with ample means, have seen nothing of it; others regard the division as variable. It is sometimes described as a narrow black line; while others refer to it as a faint pencil-like shading, and not a real division at all. One observer occasionally sees it with considerable distinctness at the very same time that another observer, with a more powerful telescope, cannot glimpse it though looking specially for such an appearance! It is difficult to reconcile such discordant experiences, and unsafe to accept results of such a contradictory nature.

The " Crape"-Ring.—A far more certain feature was discovered in the autumn of 1850 *, and one in reference to which there is unanimity of testimony. On Nov. 11 G. P. Bond, in America, and Dawes, in England, on Nov. 25, saw a nebulosity or faint luminous appearance like twilight, fringing the interior margin of the inner ring. Later observations showed this to be occasioned by a transparent ring situated immediately within the inner luminous ring. Dawes considered the new ring to be divided into two parts; but Lassell, with his large reflector, subsequently negatived this supposition. Both limbs

* Galle, at Berlin, had, twelve years previously, made an observation which, if it had been interpreted correctly, would have given him priority. In June 1838 he remarked, on several nights, that the inner boundary of the inner ring was very indistinct and "gradually lost itself towards the body of the planet." The space between the ring and Saturn was half filled with a dim veil, extending inwards from the ring. These observations failed to attract the notice their importance deserved, and Galle himself did not appreciate their full significance until the announcements of Bond and Dawes in 1850.

of Saturn may be readily perceived through the transparent ring where it crosses the globe of the planet. Some irregularities have been suspected in it at different times by various observers. In 1887 dark condensations were reported to disturb its normal aspect; but these were not seen at many observatories where such features, if real, could hardly have escaped detection.

It is strange to reflect that this transparent ring avoided discovery for so long a period. It forms a feature distinctly

Fig. 40.

Saturn, as observed by F. Terby, February 1887.

to be recognized in relatively small telescopes—in fact, Grover has seen it, where it crosses the globe of Saturn, with only 2 inches of aperture. Yet, though ever on the alert to detect new formations, and exercising constant vigilance in their pursuit, Sir W. Herschel, Schröter, and many others allowed this ring to escape them! There is no reason to suppose that it is variable, and that it was not so plain a century ago as now. It affords another instance of how easily an unknown object may elude recognition, though everyone sees it readily enough when attention is called to it.

In March 1889 a white spot was detected on the rings by Dr. Terby, at Louvain, and it was seen by other observers with comparatively small instruments. The spot was stationary, and placed near the apparent junction of the globe and rings, in the E. ansa. But with large telescopes nothing of this object could be detected: it was shown to be an optical effect.

Discordant Observations.—It is curious that the details of Saturn have occasioned more dissension amongst observers than those of any other planet. This may have partly arisen from the great distance of Saturn, the comparative feebleness of his light, and complexity of his structure. The planet is usually better defined than either Mars or Jupiter; but with tolerably high powers on small instruments the image is faint, and the features so diluted that the impressions received cannot always be depended upon, especially when the air is unsteady. A fluttering condition of the object is sufficient in itself to cause deception. Prof. Hall, in speaking of the work done by the 25·8-inch refractor at Washington in 1883, says :—" Saturn's ring has been observed, but many of the strange phenomena noted by other observers have not been seen even on the best nights." The evidence afforded by this large instrument may not always be conclusive, but in this case there can be no doubt it properly failed to show " phenomena " which had no existence.

Eccentric Position of the Rings.—The rings are slightly eccentric with regard to the ball ; in other words, the ball is not situated in the centre of the rings. Differences have been observed denoting this, though the observations are not altogether satisfactory. It has been shown theoretically that the eccentricity referred to is necessary to maintain the stable equilibrium of the system ; for were the rings perfectly concentric with the planet, they must coalesce with the ball. The preservation of so complicated a structure must evidently require judicious and nicely balanced conditions.

With the great 23-inch refractor at Princeton, U.S.A., the ball of Saturn was seen through the division in the ring in November 1883—an observation which had previously been made by Lassell in 1852.

Aspect of the Rings.—In different years the rings present a varying outline, owing to the fact of their inclination (28° 10') and to changes in the relative positions of the Earth and Saturn. At intervals of about fifteen years the rings are widely open, as they were in 1855, 1869, and 1885, and will be in 1899. At similar intervals they are rendered invisible, being turned edgeways to the Earth, as in 1848, 1862, 1877, and 1891. Since 1877 the S. side of the rings has been presented to terrestrial observers; but in 1893 the N. side will come under inspection, and remain in view until 1907. The S. side of the rings is obviously more favourably visible to observers in England and other N. latitudes, because the planet is always above the equator and attains a fair altitude when it is presented. The N. side of the rings is exposed when Saturn is in S. declination, and therefore more liable to our atmospheric disturbances owing to his comparatively low altitude. The extreme narrowness of the rings is apparent at the periods when the planet crosses the node and they are situated in the plane of the line of sight. In small telescopes they become invisible, and the finest instruments only exhibit them as thread-like extensions from the equator of the planet. Sir J. Herschel says that on April 29, 1833, the disappearance of the ring was *complete* when observed with a reflector of 18 inches aperture and 20 feet focal length. It remained visible in 1862 as a broken line of light. At such times the satellites are seen as bright beads, threading their way along the narrow wavering line of the belts. Inequalities have been observed at such times; for the line of light into which the rings are then resolved is not uniform in breadth, but appears broken and undulatory, as though indicating a very rugged character of surface.

Sir J. Herschel estimated the thickness of the rings as 250 miles, but Bond thought it far less—about 40 miles. There are great obstacles in the way of ascertaining the exact proportions of a structure so distant and offering such an extremely slender form to our view.

Further Observations required.—The globe and rings of Saturn offer an encouraging prospect for additional discoveries. Though the more prominent details have already been descried, there remain other features, probably of more delicate outline and intermittent visibility, which will be glimpsed in future years. Small instruments will scarcely be competent to deal efficiently with this object: observers who can command at least a moderate grasp of light may, however, enter upon the work with every assurance of interesting results. In this, as in other sections of observational astronomy, the student will realize that in oft-repeated observation and comparison of records and drawings he acquires a familiarity with the appearance of the object which will enable him to discern more and more of its configuration, until ultimately he feels confident he has progressed as far as the utmost capacity of his instrument will permit. It is in the sedulous application of his powers that the observer will find the key to success. Partial devotion to a subject offers a prospect far less encouraging; for observations of a disconnected character are seldom valuable.

Changes are unquestionably occurring both in connection with the ball and rings of Saturn *. Some of the discrepancies between the observations published from time to time are only to be explained on this assumption. It should therefore be the aim of observers to obtain further evidence of such variations, and this may be best accomplished by assiduously watching the lineaments of the planet during the most favourable periods of each opposition. The collection of a number of reliable materials through a series of years would undoubtedly possess weight in removing some of the anomalies of past observation, and afford us a more thorough knowledge of the delicate markings.

The rotation-period of Saturn is probably not much different from that given by the atmospheric markings seen by Herschel and Hall. But additional determinations are very desirable for many reasons. The spots which are so plentiful on Mars and Jupiter have furnished observers with a valid and concise means of ascertaining the rate of axial motion of those planets. Saturn, however, has far more sparingly provided the data for such an investigation; for if we disregard Schröter's uncertain figures, we have but two values for the rotation-period. These were fortunately effected by observers of exceptional ability, and the periods may be accepted without reservation; but other independent determinations are much required. By multiplying results of this nature, we have a prolific source

---

* Struve wrote, in 1883 :—" That changes do take place in the ring-system is sufficiently proved." Trouvelot, Schiaparelli, and others have also remarked variations of a sufficiently decided character to be placed on record.

of comparison ; and comparisons, apart from being interesting, are of importance in denoting erratic results and indicating those entitled to credence. Moreover, a reliable mean value may be sometimes deduced from multiple records ; hence it becomes advisable to secure as many as possible.

The planet should be frequently examined during every opposition with the highest powers that are consistent with a perfectly distinct image ; and the observer should closely scan the various parts of the disk, with an endeavour to trace spots, breaks, or other irregularities in the belts. Certain inequalities of tone have been occasionally apparent in past years, and they will doubtless reappear. The recovery of these features will form a welcome addition to our know-ledge, and, if adequately observed, will enable the rotation-period of the planet to be rediscussed. In an enquiry of this kind many observations are needful, and the longer the interval over which they extend the more accurate the results derived from them are likely to be. If a broken belt should appear on Saturn, the time of its passing the planet's central meridian should be recorded, either by measurement or careful esti-mation, and an ephemeris computed based on a rotation-period of $10\frac{1}{4}^{h}$, which is equal to a daily rate of nearly 843°. Then it should be carefully looked for on subsequent evenings at the times given in the ephemeris, and on every occasion when re-observed its time of transit should be noted as at first. As long as the break continues visible, so long ought it to be kept in view and the times of its central passages tabulated. It would be advisable in such a case to secure cooperation from other observers, as more numerous observations would be sure to accrue, so that, on the appearance of a marking such as that alluded to, the discoverer will do well to announce it imme-diately to other amateurs who are engaged upon planetary work and most likely to assist him. A white or dark spot, or any condensation on the belts, would of course serve the same purpose as a broken belt. The nature of the object is not necessarily to be considered, the main requirement being that it is one of which the longitude admits of determination. Markings on the belts, if they are ever discernible, must be watched with corresponding assiduity for traces of motion ;

and if such motion should betray itself, the object of the observer will be to ascertain its rate.

With reference to the narrow division in the outer ring, usually termed " Encke's division," astronomers would regard it as a gratifying advance could the doubts overhanging this feature be removed. Is it a real division in the ring, or simply a pencil-line of shading on the flat surface? Is it constant in place and appearance, or does it frequently exhibit changes both as to intensity and position? Judging from prior experiences, this particular object would appear to be extremely fugitive, and incapable of being assigned either a definite place or aspect. Yet the more pronounced and well-attested details of Saturn show no such vagaries: Cassini's division seems invariable. Are we therefore to surmise that the curious behaviour of Encke's division is to be referred to errors of observation arising from the effects of unsteady air upon a very delicate object? It is for future observers to answer these questions, and this will entail no ordinary effort, for the same impediments will be encountered in the future as in the past. But fortunately our science is rapidly progressive, and there is no doubt the mystery of Encke's division will find its solution before long. A powerful telescope, and a keen and continuous study of the outer ring, will enable some discriminating observer to tell us the true story of its phenomena.

Many other points in the Saturnian system require renewed attention, but some of them appear to be so doubtful as to scarcely deserve mention. Possibly the student had better commence his review of the planet without any of the bias or prejudice which former observations might occasion. But it is as well to know the true state of the case ; for the judgment of a careful observer is not likely to be warped by preconception, and of course some of the doubtful observations may be amply verified at a future time. Several of these have already been briefly referred to, and a few others may here be noted. The form of the shadow thrown on the rings from the ball has been observed of a curious shape, and M. Trouvelot supposes it to be variable and occasioned by changes on the level surface of the rings. The same observer has noticed transverse notches in the edges of the inner bright ring. Evidence of variation is not entirely wanting in regard to the chief division, and observers should notice whether it appears uniformly black, as it has been suggested that a gauze ring fills the interval. Exterior to the outer ring a faint luminosity has also been suspected, as though the phenomenon of the inner ring had its counterpart here. The colour of the belts on the ball should be ascribed by careful estimates, as many such observations may give an insight into the variations occurring. Some observers have alleged that the transparent ring of Bond and Dawes is subject to very perceptible alterations. It must be remembered, however, that the visible aspect of this exceedingly delicate structure is much affected by the condition of the atmosphere, and that the inclination of the Saturnian system must obviously introduce changes. When the inclination is considerable, the globe of the planet may be discerned through this ring with greater effect than at other times, because we have to look through a thinner stratum of its material.

The observer, in seeking to elucidate some of the anomalies of former researches, will possibly himself gain a knowledge of features not hitherto recognized. Of the real existence of these he should assure himself by many critical observations before venturing to announce them.

We have hinted that further discoveries upon Saturn may be considered as practically beyond the reach of small telescopes; but the gratifying fact remains that some of the more noteworthy of the known features are visible in glasses of little pretention as regards size. With a 2-inch refractor, power about 90, not only are the rings splendidly visible, but Cassini's division is readily glimpsed, as well as the narrow dark belt on the body of the planet. This sufficiently proves that a very small and portable instrument is capable of affording some excellent views of one of the most wonderful objects in the heavens. Grover has seen, with an aperture similar to that named, not only the belts and the shadow of the ball on the rings, but two of the satellites as well; and others may be equally successful.

Occultations of Saturn by the Moon.—Phenomena of this

kind were well observed in England on May 8, 1859, April 20 and Sept. 30, 1870. Those of 1859 and Sept. 30, 1870, were observed by the Rev. S. J. Johnson, who noted that "the dull hue of the planet contrasted strikingly with the brilliant yellow of the Moon." Dawes witnessed the occultation in 1859, and saw the opaque edge of our satellite sharply defined on the ball and rings of Saturn, without the slightest distortion of form. No dark shading was remarked by him contiguous to the Moon's bright edge at the reappearance, such as he and others had observed on Jupiter on the occasion of his occultation, Jan. 2, 1857. Saturn was described as of a pale greenish hue, and offered a strong contrast to the brilliant yellow lustre of the Moon. On the early morning of April 20, 1870, several observers were on the *qui vive* for this interesting occurrence; and their experiences are reported in the 'Monthly Notices R. A. S.' vol. xxx. p. 175 *et seq.*, from which the following are brief extracts :—

Mr. Ellis :—" The light of the planet, by contrast with the Moon, was very faint." Mr. Carpenter :—" There was not the least alteration in the planet's form." Capt. Noble :— " Saturn appeared of a richly-greenish yellow when compared with the brilliant white light of the Moon." Mr. G. C. Talmage :—" The difference in colour between Saturn and the Moon was most marked, the planet appearing of a yellow tint." Mr. J. Carpenter :—" At disappearance the planet was a very dull object when in contact with the Moon ; its light probably a twentieth as bright. At reappearance the planet was rather tremulous ; no distortion was noticed." On June 13, 1870, the Rev. J. Spear, of Bengal, watched the Moon pass " steadily over the planet without causing any change of form or giving any indication of the planet's light passing through an atmospheric medium. When near the Moon's limb Saturn assumed a sickly green hue."

I observed the occultation of Sept. 30, 1870, at Bristol, with a 4¼-inch refractor; but the event offered no novel traits, the most prominent feature being the difference of brightness in the Moon and Saturn. Mr. C. L. Prince observed this event with a Tulley refractor of 6·8 inches aperture, power 250. He says there was not the slightest distortion of either body, but he noticed that "the edge of the ring lingered somewhat upon the Moon's limb about the time of disappearance."

Another occultation occurred soon after new Moon on April 9, 1883, and one of the observers, Mr. Loomis, de-.scribed the disappearance of the rings as a spectacle of great interest, and said the impression was forcibly conveyed to his mind that the Moon was very much nearer to the eye than Saturn.

The Satellites.—The discovery of the eight moons of this planet ranged over the long period of 193 years. Five different observers share the honours between them. Our knowledge of the Saturnian satellites may almost be said to furnish us with a history of improvements in the telescope; for they were severally detected at epochs corresponding to instrumental advances. The following are the periods, distances, &c. of the satellites:—

No. and Name.	Mean Distance.		Sidereal Period.	Real Diam.	Date of Discovery.	Discoverer.
	Diameters of Saturn.	Miles.				
			d h .m	miles.		
7th. Mimas	1·53	115,000	0 22 37	1000	1789, Sept. 17.	W. Herschel.
6th. Enceladus...	1·97	148,000	1 8 53	......:	1789, Aug. 28.	W. Herschel *.
5th. Tethys	2·44	183,000	1 21 18	500	1684, Mar. 21.	J. D. Cassini.
4th. Dione	3·12	234,000	2 17 41	500	1684, Mar. 21.	J. D. Cassini.
3rd. Rhea	4·36	327,000	4 12 25	1200	1672, Dec. 23.	J. D. Cassini.
1st. Titan	10·12	759,000	15 22 41	3300	1655, Mar. 25.	C. Huygens.
8th. Hyperion ...	12·23	917,000	21 7 7	......	1848, Sept. 19.	Bond & Lassell.
2nd. Iapetus......	29·61	2,221,000	79 7 53	1800	1671, Oct. 25.	J. D. Cassini.

The numbers in the first column refer to the order of discovery.

Titan is by far the largest satellite, being equal to a star of the 8th mag. and visible in any small telescope. Iapetus

---

* Herschel remarks that he saw this satellite in his 20-foot speculum two years before, viz. on Aug. 19, 1787, but he was then much engaged in observations of the satellites of Uranus.

ranks next, ordinarily about 9th mag., but there are variations at different parts of the orbit similar to the variations which affect the satellites of Jupiter ; a variegated surface, and the effects of rotation, originate the changes observed and give strong support to the inference that this satellite rotates in the same period that it revolves round its primary. Tethys, Dione, and Rhea are fainter, and the difficulty of seeing them

Fig. 41.

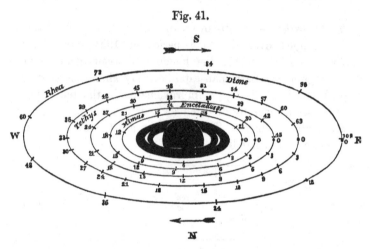

Apparent Orbits of the Five Inner Satellites of Saturn, as seen in
an Inverting Telescope.

(The arrows in the diagram show the direction of the motion of the satellites. The figures indicate the interval, in hours, from the time of last East elongation.)

is intensified by their proximity to the planet ; but a good 4-inch refractor will reveal them on a clear dark night. The others are objects for powerful instruments and pellucid skies ; but Enceladus is sometimes seen with moderate aperture. The planet being usually much inclined, his satellites are dispersed round about the rings, and are not easy of identification. Minute stars lying near the path of Saturn are very liable to be mistaken for them. But the ephemerides drawn up by Mr. Marth, and published annually by the Royal Astronomical Society, are of the utmost service to amateurs engaged in these observations. By simple reference they may readily identify the individual satellites on any night; and these ephemerides are additionally useful as giving the times of conjunctions of some of the satellites with the ends of the ring and N. and S. points of the ball.

When the thin side of the rings is presented to the Earth, transits and other phenomena may be observed in connection with the Saturnian moons ; but they appear to have been rarely recorded. Sir W. Herschel describes a " beautiful observation of the transit of the shadow of Titan over the disk in 1789, November 2." It was also seen in 1833 and 1862. The late Mr. Capron re-observed it on Dec. 10, 1877, with a 8¼-inch reflector, power 144, and made the following sketch :—

Fig. 42.

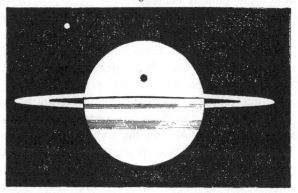

These shadow-transits admit of easy observation with appliances of very moderate capacity. Mr. Banks witnessed a phenomenon of the kind with a refractor of only 2⅞ inches, and says it was watched with the same facility and ease as the shadow of Sat. I. on Jupiter.

In looking for Iapetus it must be remembered that it is commonly situated at a great distance from the planet. Titan is relatively much nearer, and will always be recognized without trouble. Enceladus, Tethys, Dione, and Rhea hover near the outskirts of the ring ; while Mimas is extremely close to it.

Prof. Hall, with the great Washington refractor, has effected many valuable measures of this system in recent years.

He finds the orbits of the five inner satellites are sensibly circular, and that they are situated in the plane of the rings. Hyperion revolves in a very eccentric orbit, and this satellite may approach very near to Titan. He obtained an observation on March 25, 1885, which seems pertinent to the question of variation in the light of the satellites. He says:—" Mimas was remarkably bright, and could not be missed even when the full light of the planet was admitted to the eye. Generally this satellite is a difficult object, and from the ease with which it is occasionally seen one might think it variable ; but I think the difference is due to the quality of the image." There is no doubt that this is the main cause of many assumed changes in celestial objects, and especially in regard to those of a minute and delicate character.

Occultations of Stars.—Stars are rarely observed to be occulted by Saturn. Webb mentions that, in 1707 or 1708, Dr. Clark noticed a star in the interval between the ball and rings ; and Dawes once remarked a star of 8·5 mag. disappear behind the outer edge of the exterior ring. It would be extremely interesting to watch a tolerably conspicuous star pass centrally behind the Saturnian system, and to trace it through Cassini's division and the transparent ring, noting any changes in magnitude or appearance as they occurred.

CHAPTER XIII.

URANUS AND NEPTUNE.

Discovery of Uranus.—Mistaken for a Comet.—True character revealed.
—Period &c.—Observations.—Belts on Uranus.—Further Observations
required.—The Satellites.—Discovery of Neptune.—The planet observed
in 1795.—Period &c.—Observations.—Supposed Ring.—Satellite.—A
trans-Neptunian Planet.—Planetary Conjunctions.

Discovery.—While Sir W. Herschel was a musician at Bath
he formed the design of making a telescopic survey of the
heavens. When engaged in this he accidentally effected a
discovery of great importance, for on the night of March 13,
1781, an object entered the field of his 6·3-inch reflector which
ultimately proved to be a new major planet of our system.
The acute eye of Herschel, directly it alighted upon the
strange body, recognized it as one of unusual character, for
it had a perceptible disk, and could be neither fixed star nor
nebula. He afterwards found the object to be in motion, and
its appearance being " hazy and ill-defined " with very high
powers he was led to regard it as a comet, and communicated
his discovery to the Royal Society at its meeting on April 26,
1781. His paper begins as follows :—

" On Tuesday, March 13, 1781, between 10 and 11 in the
evening, while I was examining the small stars in the neighbourhood of H Geminorum, I perceived one that appeared
visibly larger than the rest. Being struck with its uncommon
magnitude, I compared it to H Geminorum and the small star
in the quartile between Auriga and Gemini, and finding it so
much larger than either of them suspected it to be a comet.
. . . . The power I had on when I first saw the comet
was 227."

The supposed " comet " soon came under the observation of
others, including Maskelyne the Astronomer Royal, and
Messier, the " Comet Ferret " of Paris. The latter, in a
letter to Herschel, said :—" Nothing was more difficult than

to catch it, and I cannot conceive how you could have hit this star or comet several times, for it was absolutely necessary for me to observe it for several days in succession before I could perceive that it was in motion."

True character revealed.—As observations began to accumulate it was seen that a parabolic orbit failed to accomodate them. Ultimately the secret was revealed. The only orbit to represent the motion of the new body was found to be an approximately circular one situated far outside the path of Saturn, and the inference became irresistible that the supposed "comet" must in reality be a new primary planet revolving on the outskirts of the solar system. This conclusion was justified by facts of a convincing nature, and its announcement created no small excitement in the scientific world. Every telescope was directed to that part of the firmament which contained the new orb, and its pale blue disk, wrapped in tiny proportions, was viewed again and again with all the delight that so great a novelty could inspire. From the earliest period of ancient history, no discovery of the same kind had been effected. The Chaldæans were acquainted with five major planets, in addition to the Earth, and the number had remained constant until the vigilant eye of Herschel enlarged our knowledge, and Saturn was relieved as the sentinel planet going his rounds on the distant frontiers of our system.

When the elements of the new body had been computed a search was instituted amongst the records of previous observers, and it was found that Herschel's planet had been seen on many occasions, but it had invariably been mistaken for a fixed star. Flamsteed observed it on six occasions between 1690 and 1715, while Le Monnier saw it on 12 nights in the years from 1750 to 1771, and it seems to have been pure carelessness on the part of the latter which prevented him from anticipating Herschel in one of the greatest discoveries of modern times.

The name Uranus was applied to the new planet, though the discoverer himself called it the *Georgium Sidus,* and there were others who termed it "Herschel," in honour of the man through whose sagacity it had been revealed.

Period &c.—Uranus revolves round the Sun in 30,687 days, which very slightly exceeds 84 terrestrial years. His mean distance from the Sun is 1,782,000,000 miles, but the interval varies between 1,699 and 1,865 millions of miles. The apparent diameter of the planet undergoes little variation ; the mean is 3″·6, but observers differ. His real diameter is approximately 31,000 miles, and the polar compression about $\frac{1}{13}$, though this value is not that found by all authorities.

Observations.—The planet near opposition shines like a star of the 6th magnitude, and is observable with the naked eye. He emits a bluish light. While engaged in meteoric observations, I have sometimes followed the planet with the naked eye during several months, and noted the changes in his position relatively to the stars near. It is clear from this that Uranus admitted of detection before the invention of the telescope.

A luminous ring, similar to that of Saturn, was at first supposed to surround Uranus, and Herschel suspected the existence of such a feature on several occasions; but it scarcely survived his later researches, and modern observations have finally disposed of it.

Lassell, when working with his 2-foot speculum at Malta, thought he saw a spot near the centre of the planet's disk, but he considered this might possibly be due to an optical illusion. In 1862, Jan. 29, he said :—" I received an impression which I am unable to render certain of an equatoreal dark belt." In the early months of 1870, Mr. Buffham, using a 9-inch " With " mirror, powers 212 and 320, saw bright spots and zones on the planet, and inferred a rotation-period of about 12 hours. On Jan. 16, 1873, when definition was very good, no traces of any markings were visible in Lord Rosse's 6-foot reflector. In May and June 1883 Prof. Young, having the advantage of the fine 23-inch refractor at the Princeton Observatory, observed two faint belts, one on each side of the equator, and much like the belts of Saturn. On March 18, 1884, Messrs. Thollon and Perrotin, with the 14-inch equatoreal at Nice, remarked dark spots similar to those on Mars, towards the centre of the disk, and a white spot was seen on the limb. Two different tints were perceived, the colour of the N.W. hemisphere being dark, and that of the S.E. a bluish-white colour. In April observations were continued, and the white spot was seen " rather as a luminous band than a simple spot," but it was most conspicuous near the limb. The observers thought the appearances indicated a rotation-period of about 10 hours. The brothers Henry at Paris, in 1884, invariably noticed two belts lying parallel to each other, and including between them the brighter equatoreal zone of the planet. Their results apparently show that the angle between the plane of the Uranian equator and that of the satellite-orbits is about 41°.

M. Perrotin, with the great 30-inch equatoreal at Nice, re-observed the belts in May and June 1889. He wrote that dark parallel bands were noticed several times, and they were very similar to the belts of Jupiter. On May 31 and June 1

Fig. 43.

Uranus and his Belts. 1884.

and 7 the direction of the Uranian belts was measured, and the mean result showed |that the plane of the equator of Uranus differs little (about 10°) from the common plane of the orbits of the satellites. This deduction is not, it will be observed, consistent with that of the Brothers Henry at Paris, who found a difference of 41°. M. Perrotin notes that the bands of Uranus do not always present the same aspect.

They vary in size and number in different parts of their circumference. This unequal distribution raises the hope that by an attentive study of these bands it will be possible to determine the duration of the planet's rotation.

Further Observations required.—In the case of an object so faint and diminutive as Uranus, a powerful telescope is absolutely required to deal with it effectively. A small instrument will readily show the disk, and present the picture that caught the eye of Herschel more than a century ago, but considerable light and power must be at command if the observer would enter upon a study of the planet's surface-markings. With my 10-inch reflector I have suspected the existence of the belts, but under high powers the image is too feeble to exhibit delicate forms of this character. It is to be hoped that with the large telescopes now available at various observatories, some attention will be given to this planet, more especially with regard to the study of the belts and determination of the rotation-period. Amateurs will have little trouble in picking up Uranus ; his position can be learnt from an ephemeris and marked upon a star-map. A little careful sweeping with a low power in the region indicated will soon reveal the object sought for, and a higher power may then be applied to expand the disk and render identification certain.

It may be mentioned as an interesting point that some fifty years after the discovery of Uranus by Sir W. Herschel the planet was accidentally rediscovered by his son Sir John Herschel, who mentioned the fact as follows in a letter to Admiral Smyth, written on Aug. 8, 1830 :—" I have just completed two 20-foot reflectors, and have got some interesting observations of the satellites of Uranus. The first sweep I made with my new mirror I *rediscovered* this planet by its *disk*, having blundered upon it by the merest accident for 19 Capricorni." Had the father failed to detect this planet in 1781, the discovery might therefore have been made by the son half a century later.

Some spectroscopic observations of Uranus made in 1889 with Mr. Common's 5-foot reflector, appear to show that the planet " is to a large extent self-luminous." But Mr. Huggins on June 3 seems to have obtained a different result (see ' Monthly Notices,' xlix. p. 404 *et seq.*).

The Satellites.—For many years it was supposed that Uranus possessed six satellites, all of which were discovered by Sir W. Herschel, but later observations proved that four of these had no existence. They were small stars near the planet. But two of Herschel's satellites were fully corroborated, and two new ones were discovered by Lassell and Struve. The number of known satellites attending Uranus is four, and it is probable that many others exist, though they are too minute to be distinguished in the most powerful instruments hitherto constructed. The following are the periods, distances, &c., of the known satellites :—

Number and name.	Mean distance.		Max. Elongation.	Date of Discovery.	Discoverer.
	Diameters of Uranus.	Miles.			
3rd. Ariel	4·03	125,000	$1\frac{v}{2}$	1847, Sept. 14.	W. Lassell.
4th. Umbriel ...	5·61	174,000	15	1847, Oct. 8.	O. Struve.
1st. Titania ...	9·19	285,000	33	1787, Jan. 11.	W. Herschel.
2nd. Oberon ...	12·32	382,000	44	1787, Jan. 11.	W. Herschel.

Titania and Oberon are the two brightest satellites, but none of them can be seen except in large instruments. The two outer ones are said to have been glimpsed in a 4·3-inch refractor, but this feat is phenomenal, and certainly no criterion of ordinary capacity. Sir J. Herschel found them tolerably conspicuous in a reflector of 18 or 20 inches aperture, and mentioned a test-object by which observers might determine whether their telescopes were adequate to reveal them. This test is a minute double star lying between the stars β' and β^2 Capricorni. The magnitudes are 15 and 16, and distance 3″. Relatively to the satellites of Uranus this faint double is a " splendid object."

From observations with large modern instruments it appears highly probable that the four known satellites must

be considerably larger than any others which may be revolving round the planet.   A curious fact in connection with
these satellites is that their motions are retrograde.

Fig. 44.

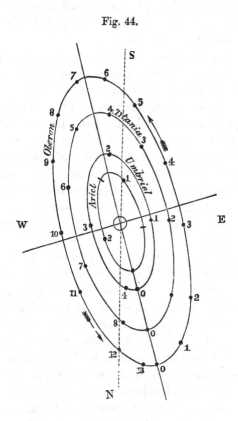

Apparent Orbits of the Satellites of Uranus, as seen in an
Inverting Telescope.

(The small circle in the above diagram represents the planet and is on the
same scale as the orbits. The arrows show the direction of the motion
of the satellites, and the figures indicate the number of days from the
time of the last North elongation.)

Discovery of Neptune.—The leading incidents in the nar-
rative of the discovery of Uranus and Neptune present a
great dissimilarity—Uranus was discovered by accident,
Neptune by design. Telescopic power revealed the former,

while theory disclosed the latter. In one case optical appliances afforded the direct means of success, while in the other the unerring precision of mathematical analysis attained it. The telescope played but a secondary part in the discovery of Neptune, for this instrument was employed simply to realize or confirm what theory had proven.

Certain irregularities in the motion of Uranus could not be explained but on the assumption of an undetected planet situated outside the known boundaries of the system. Two able geometers applied themselves to study the problem of these irregularities, and to deduce from them the place of the disturbing body. This was effected independently by Messrs. Le Verrier and Adams ; and Dr. Galle, of Berlin, having received from Le Verrier the leading results of his computations, and the intimation that the longitude of the suspected planet was then 326°, found it with his telescope on the night of Sept. 23, 1846, in longitude 326° 52'. The calculated place by Prof. Adams was 329° 19' for the same date and less accurate than the prediction of Le Verrier. The former had priority both in attacking the problem and resolving it, though unfortunately his efforts were not backed up in a practical way. But for the supineness of certain officials, there is little doubt that the planet would have been telescopically discovered in the autumn of 1845, when it was within 1° 49' of the place attributed to it by Prof. Adams. Delays occurred owing to the doubts prevailing, and in the meantime the planet was found elsewhere. This circumstance does not rob Prof. Adams of his hard-earned laurels, though it shows how seriously official negligence can mar the character of a discovery.

Observations in 1795.—The name given to the new planet was Neptune. When the elements were computed it was found that they presented rather large differences with those theoretically computed by Messrs. Le Verrier and Adams. It was also found that the planet had been previously observed by Lalande on May 8 and 10, 1795, but its true character escaped detection. This astronomer had observed a star of the 8th mag. on May 8 ; but on May 10, not finding the same star in the exact place noted on the former evening, he rejected the first observation as inaccurate and adopted the second, marking it doubtful. Had Lalande exercised a little discretion, and confided in his work, he would hardly have allowed the matter to rest here. A subsequent observation would at once have exhibited the cause of the discrepancy, and the mathematical triumph of Le Verrier and Adams, half a century later, would have been forestalled. Lalande, like Le Monnier, the unsuspecting observer of Uranus, let a valuable discovery slip through his hands.

Period &c.—Neptune revolves round the Sun in 60,126 days, which is equal to rather more than 164½ of our years. His mean distance from the Sun is 2,792,000,000 miles, and his usual diameter 2‴·7. He exceeds Uranus in dimensions, his real diameter being 37,000 miles.

Observations.—Our knowledge of this distant orb is extremely limited, owing to his apparently diminutive size and feebleness. No markings have ever been sighted on his miniature disk, and we can expect to learn nothing until one of the large telescopes is employed in the work. No doubt this planet exhibits the same belted appearance as that of Uranus, and there is every probability that he possesses a numerous retinue of satellites. In dealing with an object like this small instruments are useless; they will display the disk, and enable us to identify the object and determine its position if necessary, but beyond this their powers are restricted by want of light.

Supposed Ring.—Directly the new planet was discovered, Mr. Lassell turned his large reflector upon it and sought to learn something of its appearance, and possibly detect one or more of its satellites. On October 3 and 10, 1846, he was struck with the appearance of the disk, which was obviously not perfectly spherical. He subsequently confirmed this impression, and concluded that a ring, inclined about 70°, surrounded the planet. Prof. Challis supported this view, but later observations in a purer sky led Mr. Lassell to abandon the idea. Thus the ring of Neptune, like the ring of Uranus, though apparently obvious at first, vanished in the light of more modern researches.

The Satellite.—But if Mr. Lassell quite failed to demonstrate the existence of a ring, he nevertheless succeeded in discovering a satellite belonging to the planet. This was on Oct. 10, 1846. The new satellite was found to have a period of 5^d 21^h 3^m, and to be situated about 220,000 miles distant from the planet. Its apparent star mag. is 14, and at max.

Fig. 45.

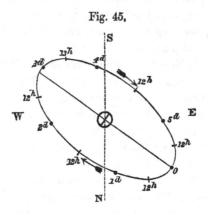

Apparent Orbit of the Satellite of Neptune, as seen in an
Inverting Telescope.

(The small circle in the above diagram represents the planet, the arrows
show the direction of motion, and the figures indicate the interval
from the time of last North-east elongation.)

elongation it extends its excursions to $18''$ on either side of its primary. Compared with the other satellites of our system the one attending Neptune must be excessive in regard to size, or it could not be discerned at the vast distance separating it from the Earth.

A trans-Neptunian Planet.—Is there a planet beyond Neptune? Prof. Forbes wrote a memoir in 1880 tending to prove that two such planets exist. From the influences exerted by these bodies on certain comets of long period, he approximately deduced the positions of the former, and they were searched for with the great Washington refractor, but without success. Flammarion and Todd have also arrived at conclusions affirming the existence of a planet outside Neptune; but the idea has not yet been realized by its telescopic discovery.

Planetary Conjunctions.—Before concluding this chapter, an allusion should be made to a noteworthy class of events, viz., planetary conjunctions. These include some of the most attractive aspects displayed by the heavenly bodies, and they are sometimes witnessed by ordinary persons with the same amount of gratification as by the astronomical amateur. In almanacks the times of such conjunctions are given, so that intending observers may always be prepared for these events. In a strict sense a conjunction occurs at the instant when two or more bodies have the same right ascension, but the term is here intended to have a more general reference, *i. e.*, to denote the assembling together of two or more planets in the same region of the firmament. Historical records furnish us with a considerable number of planetary conjunctions, and some of them were attentively observed long before the telescope came into use. Thus in 2012 B.C., Feb. 26, the Moon, Mercury, Venus, Jupiter, and Saturn were in the same constellation, and within 14° of one another. In 1186 A.D., Sep. 14, the Sun, Moon, and all the known planets are said to have been situated in Libra. In 1524 Venus, Mars, Jupiter, and Saturn were near together. Many similar instances might be quoted, but this is unnecessary. Occasionally the conjunctions were so close that one planet appeared to occult another. Kepler refers to an occultation of Jupiter by Mars which he saw on January 9, 1591; but this would really be a transit of Mars across the disk of Jupiter, if contact actually occurred, for the apparent diameter of Jupiter always exceeds that of Mars. Mœstlin seems to have witnessed an occultation of Mars by Venus on Oct. 3, 1590. It is probable, however, that these were near approaches only. A genuine occultation of Mercury by Venus was telescopically observed on May 17, 1737.

On the evening of March 3, 1881, the new Moon, Venus, Jupiter, and Saturn formed a brilliant quartet in Pisces. On the morning of July 21, 1881, I saw the Moon, Venus, Mars, Jupiter, Saturn, and Aldebaran in the same region above the eastern horizon. There was a very close conjunction of Mars and Saturn on the morning of Sept. 20, 1889. Mr. Marth computed that the nearest approach would occur at $8^h 7^m$ A.M., when the distance between the centres would be 54″·8 and less than that (74″) observed at the time of the close conjunction of the same planets on June 30, 1879.

The interest centred in the conjunction of Sept. 20, 1889, was enhanced by the fact that Regulus was only 47′ distant,

Fig. 46.

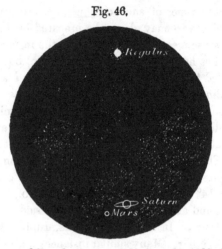

Mars, Saturn, and Regulus in same field, Sept. 20 1889, 4ʰ 45ᵐ A.M.

while Venus was also in the same region. I observed this phenomenon in my 10-inch reflector, and with the help of a comet-eyepiece made the above sketch of the positions of the objects as they were presented in the field.

Perhaps there is not much scientific importance attached to the observation of these conjunctions, though comparisons of colour and surface-brilliancy are feasible at such epochs, and are not wholly without value. As spectacles merely, they possess a high degree of interest to everyone who " considers the heavens."

CHAPTER XIV.

COMETS AND COMET-SEEKING.

Ideas concerning Comets.—Appearance.—Large number visible.—Nature of Apparition.—Tenuity of Comets.—Differences of Orbit.—Discoveries of Comets.—Large Comets.—Periodical Comets.—The Comets of Halley, Encke, Biela, Brorsen, Faye, D'Arrest, Pons-Winnecke, and Tuttle.—Grouping.—Further Observations required.—Nomenclature of Comets.—Curiosities of Comets.—Naked-eye Comets.—Comet-seeking. —English weather.—Aperture and Power required.—Annual rate of Discovery.—Telescopic Comets and Nebulæ.—Ascertaining Positions.—Dr. Doberck's hints.—Prizes.

SUPERSTITIOUS ideas with regard to comets as the harbingers of disaster have long since been discarded for more rational opinions. They are no longer looked upon as ill-omened presages of evil, or as

> " From Saturnius sent,
> To fright the nations with a dire portent."

Many references are to be found among old writings to the supposed evil influence of these bodies, and to the dread which their appearance formerly incited in the popular mind. Shakespeare makes an allusion to the common belief :—

> " Hung be the heavens with black, yield day to night !
> Comets, importing change of time and states,
> Brandish your crystal tresses in the sky ; "

and in relation to the habit of connecting historical events with their apparition, he further says :—

> " When beggars die, there are no comets seen ;
> The heavens themselves blaze forth the death of princes."

But, happily, the notions prevalent in former times have been superseded by the more enlightened views naturally resulting from the acquirement and diffusion of knowledge ; so that comets, though still surrounded by a good deal of mystery, are now regarded with considerable interest, and welcomed, not only as objects devoid of malevolent character, but as furnishing many useful materials for study. Mere superstition has been put aside as an impediment to real progress, and a more intelligent age has recognized the necessity of dealing only with *facts* and explaining them according to the laws of nature ; for it is on facts, and their just interpretation, that all true searchers after knowledge must rely. Comets are properly regarded as bodies which, though far from being thoroughly understood in all the details of their physical structure and behaviour, have yet a wonderful history, and one which, could it be clearly elucidated, would unfold some new and marvellous facts. Under these circumstances we need evince no surprise that these visitors are invariably hailed with enthusiasm, not only by scientific men, who make them the special subjects of close observation, but by everyone who regards celestial "sights and signs" with occasional attention.

Appearance.—From whatever point of view a large comet is considered, it deserves all the interest manifested in it and all the labour expended in its investigation. Whilst its grand appearance in the firmament arrests the notice of all classes alike, and is the subject of much curious speculation amongst the uninformed, it merits, apart from other considerations, the most assiduous observation on account of the singular features it displays and the striking variations they undergo. Indeed, the visible deportment of a comet during its rapid career near perihelion is so extraordinary as to form a problem, the solution of which continues to defy the most ingenious theories. The remarkable changes in progress, the quickness and apparent irregularity of their development, are the immediate result of a combination of forces, the operations of which can neither be defined nor foreseen. Jets of flame and wreaths of vapour start from the brilliant nucleus; while, streaming away from the latter, in a direction opposite to the Sun, is the fan-shaped tail, often traceable over a large span of the heavens and commingling its extreme fainter limits with the star-dust in the background.

Large number visible.—The orbits of 400 comets have now been computed, and more than 500 others have been observed; so that these bodies are extremely plentiful. Kepler described them to be as numerous as the fishes in the sea, and no. doubt the allegory is justly applied. Their vagaries of form, size, and place are equally noteworthy; and those who enter upon the discussion of facts relating to these objects will find an endless store of interesting materials, opening up a wide field for conjecture.

Nature of Apparition.—The apparition of a comet may be either gradual or sudden. Usually the telescope gives us the earliest intimation that one of these bodies is approaching us*. It is first seen as a small round nebulosity, with probably a central condensation or stellar nucleus of the 10th or 11th mag. The whole object brightens and expands as its distance grows less, and it assumes an elongated form preparatory to the formation of a tail. The latter varies greatly in different instances: it may either be a narrow ray, as shown in the southern comet of January 1887, or a fan-shaped extension like that of the great comet of 1744. Barnard's Comet of December 1886 exhibited a duple tail. Occasionally a fine comet bursts upon us suddenly, like that of 1843 or 1861. The former was sufficiently bright to be discovered when only 4° from th Sun, and the latter presented itself quite unexpectedly as a magnificent object even in the strong twilight of a June sky.

Tenuity of Comets.—Comets are noteworthy for the extreme thinness of their material. The smallest stars may be discerned through the denser portions of the head, without suffering any apparent diminution of light. Yet such stars would be quite obscured by the interposition of a minute speck of cloud or by a little fog or any vapour of trifling density. Comets are visible in the form of transparent nebulosities; and their mass must be inconceivably small relatively to the enormous space over which they frequently extend. Sir J. Herschel has described the "all but spiritual texture" of comets; and other authorities have referred to them as feeble wreaths of vapour, which, though obeying the laws of gravitation and suffering

* Donati's Comet of 1858 and Coggia's Comet of 1874 may be mentioned as good examples of the gradual approach and development of these visitors witnessed by means of the telescope.

much perturbation, are yet themselves incapable of exercising any disturbing influence upon the other bodies near which they pass. It has been asserted that comets would show phases were they rendered luminous by reflected sunlight, and that, such features being absent, these bodies must possess a phosphorescence of their own sufficient to cause the glow observed. This idea, however, is hardly consistent with our present knowledge. Comets are not compact and coherent masses of matter ; they more likely represent vast groups of planetary atoms, more or less loosely dispersed and sometimes forming streams. The effect of sunshine upon such assemblages will be that the whole mass becomes illumined according to density, and that no phase will be apparent, inasmuch as the light is enabled to penetrate through its entirety.

Differences of Orbit.—When three trustworthy observations of a comet's place have been made, its orbit may be computed. This may be either an ellipse, a parabola, or hyperbola. If an *ellipse* the comet is periodical, and the period depends upon the degree of eccentricity. If a *parabola* the comet will not be seen again, because this form of orbit does not reunite; it consists of branches equally divergent and uniting at perihelion, but extending outwards indefinitely in nearly parallel lines and without convergence. If a *hyperbola* the comet is also not returnable ; the branches of the orbit are widely divergent, and show no tendency to parallelism. These several forms of orbit are somewhat different as applied to various comets, but they are the same in effect. Thus Tempel's Comet of 1867 revolves in an ellipse having an eccentricity of about 0·4630, while that of Halley's Comet is 0·9674. No doubt some of the parabolic orbits applied to comets really represent very eccentric ellipses; but the parabola is a convenient form of orbit for computation, and unless ellipticity is very decided it indicates the path with sufficient accuracy.

Discoveries of Comets.—In the latter part of the last century Messier, Mechain, and Miss Herschel shared nearly all the cometary discoveries between them. Then Pons entered the field, and he may be said to have monopolized this branch during the period from 1802 to 1827, for he was the first to announce thirty comets. Pons died in 1831, but the search was actively continued by others. In about 1843 a great rise became apparent in the rate of these discoveries ; and we find Di Vico, Mauvais, and Brorsen very successful at this period. Later on, the work was sustained with the same prolific results by Klinkerfues, Bruhns, and Donati, and subsequently by Winnecke, Tempel, and Coggia. Swift and Borrelly also assisted materially to swell our knowledge ; while during the last few years Barnard and Brooks have exhibited a surprising amount of zeal in this department. Since 1881 no less than twenty-six comets are to be enumerated as the fruits of their endeavours, and they are still engaged in nightly explorations of the sky with similar ends in view. Their diligent pursuit of these fugitive bodies will doubtless result in many further additions during ensuing years.

It is a curious circumstance that Sir W. Herschel, during all his star-gaugings and sweeps for nebulæ, never discovered a comet. He found a nebula on Dec. 18, 1783, near δ Ceti, which he described as "small and cometic." In Sir J. Herschel's 'General Catalogue of Nebulæ,' 1864, p. 17, this object is presumed to have been a comet, as it could not be identified; but at p. 45 the doubts are cleared up, and Sir W. Herschel's nebula, the position of which was only roughly given, is shown to be the same as another very near ; it is No. 1055 of the new 'General Catalogue' published by the Royal Astronomical Society in January 1888. Quite possibly Sir W. Herschel's lists of nebulæ contain several comets, as some of his objects are missing; but errors of observation in ascribing positions may explain this. Herschel himself, in speaking of a comet visible in the winter of 1807-8, says :—" If I had met the comet in one of my sweeps, as it appeared between Dec. 6 and Feb. 21, I should have put it down as a nebula. Perhaps my lists of nebulæ, then, contain some comets."

Large Comets.—The most widely observed and attractive

class of comets includes those of large proportions, as they are not only visible to the naked eye, but exhibit features having the lustre necessary to permit of their examination with high magnifying powers. A brief summary of some of the finest comets of modern times is subjoined; but, to save space, a few only of the more salient facts concerning them are given:—

1577, Nov. and Dec.—Observed by Tycho Brahe. At the end of November it had a double tail; the longest of the two branches was about 20°. This comet was visible in the day-time.

1618 II., Nov.—" The length of its tail equalled in extent one sixth part of the zodiac." On Nov. 18 it was estimated as 40°. Longomontanus, however, described it as 104° long, and Cysatus estimated it as 75°. Kepler referred to it as the largest comet that had appeared for a hundred and fifty years.

1680, Dec.—A fine comet, which on Dec. 12 had a narrow tail about 80° long. The nucleus was equal to a 1st mag. star. Hooke remarked jets of flame issuing from the nucleus. At perihelion the comet approached very near the Sun's surface, similarly to the fine comets of 1843, 1880, and 1882.

1744, Jan.-Feb.—Probably the largest comet of the 18th century. At one time it displayed six tails, each of which was 4° in breadth. The head was so bright that it was perceived with the naked eye in full sunshine. At the middle of February the tail was 24° long, and it was divided into two branches.

1769, Sept.—Discovered on Aug. 8 by Messier. On Aug. 30 the comet had a trifid tail; there was a central ray of 24° and two outlying ones of 4° each. On Sept. 19 the tail had increased to 75°, and a few nights later Pingre estimated it as 90° and 97°.

1811 I., Sept.-Oct.—A very fine comet. The tail was branched; it did not, however, exceed 25° in length and about 6° in breadth. Sir W. Herschel found the nucleus to be 428 miles in diameter. This remarkable comet remained visible during a period extending over seventeen months. Its period is approximately 3000 years.

1843, Mar.—Visible in the daytime. On Mar. 4 its tail was 69° in length; it was very narrow, being only 1¼° in breadth throughout. At perihelion this object passed very near to the Sun, like the great comet of 1680. It revolves in an elliptical orbit; period about 376 years. This comet swept past perihelion with a velocity of 366 miles per second! The real length of its tail was 200 millions of miles!

1858 VI., Sept.–Oct.—Donati's Comet: one of the most brilliant comets of the 19th century. Early in October it displayed a tail about 40° long, and on the 5th it passed over the star Arcturus. Its period of revolution appears to be about 2000 years.

1861 II., June–July.—Became suddenly visible at the end of June. In the opinion of Sir John Herschel this comet surpassed in grandeur the comets of 1811 and 1858. On June 30 the nucleus was equal to the brightness of Venus, and the tail was 80° long; but early in July it increased to 90°. One observer estimated its length as 100° on July 2. This comet remained visible during twelve months. It appears to have an elliptical orbit, with a period of 409 years.

1874, July.—Coggia's Comet: a fine object in the northern sky. On July 14 the tail was 35° long, and it remained visible several days after the nucleus had disappeared below the horizon. The nucleus was about equal to a star of the 1st mag. Orbit probably elliptical, with a period of about 5711 years.

1880 I., Jan.–Feb.—A southern comet, with a long narrow tail, variously estimated from 30° to 40° in length. It passed very near to the Sun, and presents an orbital resemblance to the fine comets of 1680 and 1843.

1881 III., June–July.—This large comet appeared in the northern heavens on June 22, and became generally visible to observers in England. On the 27th it had a tail 15° long. Its period of visibility extended over nine months.

1881 IV., Aug.—This comet is scarcely entitled to rank as one of exceptional character; but it was a conspicuous object for several weeks in August, and had a tail 6° long on the 19th.

1882 III., Oct.—Well visible in the morning sky, with a tail 22° long. The nucleus underwent remarkable changes, and on Oct. 23 it showed four or five bright points or nuclei, looking like "a string of beads." The comet threw off several small condensations, which were observed as separate comets near the parent mass. At perihelion this comet passed very close to the Sun, like the comets of 1680, 1843, and 1880 ; and these bodies were suspected to have an intimate relation, if not an absolute identity. But subsequent inquiries disproved this startling supposition ; for the comet of 1882 was shown to have a period of about 718 years.

1887 I., Jan.—A fine southern comet, presenting many points of resemblance to that of 1880 I. On Jan. 22, as observed at Adelaide, the comet had a long narrow tail of about 30°, but no well-defined nucleus. On the same date, at the Cape of Good Hope, the tail appeared as a narrow ribbon of light, quite straight, and of nearly uniform brightness throughout its length. It was visible in the same region of the sky as the comet 1880 I., and came into view with equal suddenness.

Periodical Comets.—On page 235 is a list of the periodical comets as at present known. Some of these, marked with an asterisk, have only been observed at one return, and therefore await complete confirmation.

Many other comets have shown indications of pursuing elliptical orbits. Amongst those of short period may be mentioned 1743 I., 1766 II., 1783 I., 1819 IV., 1844 I., and 1873 VII. The following are examples of longer periods:—

Comet.	Period.	Comet.	Period.
1862 III.	121 years.	1877 III.	28,000 years.
1857 IV.	234 „	1850 I.	29,000 „
1861 I.	415 „	1780 I.	75,314 „
1860 III.	1089 „	1844 II.	102,050 „
1889 IV.	5100 „	1744	122,683 „
1877 II.	8393 „	1849 I.	382,801 „
1847 III.	13918 „	1882 I.	400,000 „

These figures are to be regarded as approximations only.

Name.	Period, in years.	Perihelion Passage.	Long. of Perihelion.	Long. of Ascending Node.	Inclination.	Motion.	Next Return.
Encke	3·29	1888, June 28	158 36	334 39	12 53	D	1891
. Tempel (1873)	5·20	1878, Sept. 7	306 8	121 1	12 46	D	1894
* Barnard	5·40	1884, Aug. 16	301 2	5 9	5 28	D	1895
Brorsen	5·46	1879, Mar. 30	116 15	101 19	29 23	D	1895
Pons-Winnecke	5·73	1886, Sept. 16	276 4	101 56	14 27	D	1892
Tempel (1867)	5·98	1879, May 7	238 11	78 46	9 47	D	1891
Tempel-Swift	5·99	1880, Nov. 8	43 0	296 42	5 31	D	1892
* Brooks (1886)	6·30	1886, June 7	229 46	53 3	12 56	D	1892
* Spitaler	6·40	1890, Oct. 26	58 24	45 8	12 52	D	1897
Biela	6·62	1852, Sept. 23	109 8	245 52	12 33	D	?
D'Arrest	6·64	1890 September	319 9	146 9	15 43	D	1897
* Finlay	6·67	1886, Nov. 22	7 34	52 30	3 2	D	1893
** Wolf	6·78	1884, Sept. 27	352 31	206 18	25 16	D	1891
* Swift	6·91	1889, Nov. 29	69 29	331 27	10 3	D	1896
* Brooks (1889)	7·07	1889, Sept. 30	1 26	17 59	6 4	D	1896
Faye	7·57	1888, Aug. 19	50 56	209 42	11 20	D	1896
* Denning	8·69	1881, Sept. 13	312 31	65 57	6 51	D	1899
* Peters	12·80	1846, June 1	240 7	260 28	30 24	R	?
Tuttle	13·66	1885, Sept. 11	116 28	269 42	54 19	D	1899
* Tempel (1866)	33·18	1866, Jan. 11	60 28	231 26	17 18	R	1899
* Stephan	33·62	1867, Jan. 20	75 52	78 36	18 13	D	1900
* Westphal	67·77	1852, Oct. 13	43 12	346 13	40 59	D	1920
Pons	71·48	1884, Jan. 25	93 21	254 6	74 3	D	1955
Olbers	72·45	1887, Oct. 8	149 45	84 30	44 35	D	1960
* Di Vico	73·25	1846, Mar. 6	90 35	77 36	84 57	D	1919
* Brorsen	74·97	1847, Sept. 10	79 13	309 49	19 8	D	1922
Halley	76·37	1835, Nov. 15	304 32	55 10	17 45	R	1912

Halley's Comet.—A fine comet with a tail about 15° long appeared in the summer of 1682, and Halley computed the orbit according to the method explained by Newton. He then consulted observations of previous comets, and discovered a great similarity in the paths of large comets seen in 1531 and 1607 to that of the one he himself had observed in 1682. He thereupon suspected the three bodies to be one and the same, and advised posterity to maintain a strict watch for the comet's return in about 1758 or 1759. On pursuing his investigations still further, he alighted upon records of comets in 1305, 1380, and 1456, which greatly strengthened his opinion that the comet of 1682 moved in an elliptical path with a period of about 75½ years. He termed this body "the Mercury * of comets, revolving round the Sun in the smallest orbit," and said that, should it reappear according to his prediction in about the year 1758, "impartial posterity must needs allow this to be the discovery of an Englishman."

As the time drew near for the return of the comet, interest became intensified, and computations were made by Clairaut with a view to determine the precise epoch when it would arrive at perihelion. He found that the comet would be retarded by the action of Jupiter and Saturn, but that perihelion would be reached at the middle of April 1759, subject to an uncertainty of 30 days. The comet was rediscovered on Dec. 25, 1758, by Palitzch, an amateur astronomer at Politz, near Dresden, who employed a telescope of 8 feet focal length, and appears to have anticipated Messier and others who were on the alert for it. It arrived at perihelion on March 12, 1759, and within a month of the date announced by Clairaut. Early in May it had a tail nearly 50° long, and presented a fine aspect in the heavens. Thus the sagacity of Halley had revealed a periodical comet—the first known. It duly returned again in 1835, and received all the attention which a body so replete with historical associations deserved.

Encke's Comet.—Until the year 1819 Halley's Comet was

---

* It ought, perhaps, in the present state of our knowledge, to be termed "the Neptune of comets;" for it has the longest period of any comet whose path has been definitely ascertained by multiple returns to perihelion.

the only one certainly known to be periodical. Then the able deductions of Encke presented us with a veritable "Mercury* of comets." He showed that a small comet, discovered by the unwearying Pons of Marseilles on Nov. 26, 1818, was

Fig. 47.

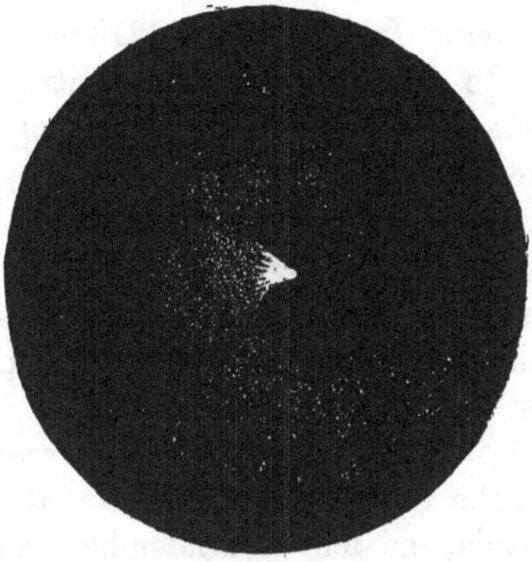

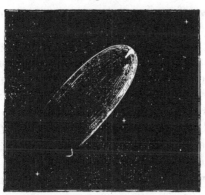

Comet 1862 III. (Aug. 19, 1862).

Fig. 48.

Sawerthal's Comet, 1888 I. (March 25, Brooks).

really identical with three previous comets—viz. 1786 I. (Mechain), 1795 (C. Herschel), and 1805 (Thulis),—and that its period was a very short one of about $3\frac{1}{3}$ years. Its return

* Encke's Comet has the shortest period of all the known comets.

to perihelion was predicted to occur on May 24, 1822, and
this was observed in the southern hemisphere. It returned
again on Sept. 16, 1825, and on this occasion the circumstances
were more favourable. Since 1825 this object has effected
nineteen returns to perihelion, and one of the most singular
facts noticed in connection with it is that its period is gradually shortening. In 1795 it was 1212 days, while in 1858
it was 1210. In order to explain this contraction of orbit, it
has been necessary to assume the existence of a thin medium
in space capable of affording a slight resistance to the tenuous
materials of a comet, though not dense enough to appreciably
affect the motions of planets. If this closing-up of the orbit
and shortening of period continue to operate through a vast
interval of time, Encke's Comet must be ultimately precipitated upon the Sun!*

   Biela's Comet.—This comet was discovered on Feb. 27,
1826, by Wilhelm von Biela, an Austrian officer, at Joseph-
stadt in Bohemia, and ten days later by Gambart at Marseilles. It was found to be revolving in an orbit of short
period, and its elements presented an agreement with those
of the comet of 1772 (Montaigne) and 1806 I. (Pons).
Identity was inferred, and the next return was fixed for
Nov. 27, 1832, when the object reappeared with great
punctuality. At the end of 1845 this comet displayed some
startling phenomena ; for it divided into two portions, apparently quite disconnected, and which travelled side by side,
separated by an interval of more than 150,000 miles! The
double comet was observed again in 1852, when the interval
separating them had, however, increased eightfold, for the
dark space between measured 1,250,000 miles. This instance
of a divided comet is by no means unique. The great comet
of 1882 underwent a process of disruption, by throwing off

---

* Newton conjectured that comets formed "the aliment by which
suns are sustained," his opinion being that the former bodies finally
coalesced with the suns round which they revolved. He remarked :—
"I cannot say when the Comet of 1680 will fall into the Sun,—possibly
after five or six revolutions ; but whenever that time shall arrive, the heat
of the Sun will be raised by it to such a point that our globe will be burnt
and all the animals upon it will perish."

small masses of nebulosity, which, however, survived the separation only a few days. Brooks's Comet (1889 V.) was found by Barnard, on Aug. 1, 1889, to be divided into four parts! Two of these had a brief existence; but one of the

Fig. 49.

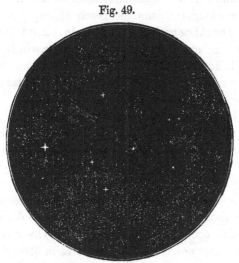

Brooks's double Comet, Sept. 17, 1889.
(10-inch reflector, power 60. W. F. Denning.)

minor fragments retained a very distinct appearance near the parent mass during the ensuing months of September and October. The phenomena of Biela's Comet found an excellent counterpart here.

Since 1852 Biela's Comet has been lost. The most assiduous observations have failed to recover it, and the conclusion seems irresistible that further disintegrations have occurred and that its material has been dispersed beyond recognition. The great meteoric storms of Nov. 27, 1872 and 1885, were derived from this comet, and there is little reason to hope that *as a comet* it will ever be seen again.

Brorsen's Comet.—A small comet was discovered in Pisces by Theodor Brorsen, at Kiel, on the evening of Feb. 26, 1846. Its observed path soon gave traces of an elliptical orbit; and the period was found to be about 5·58 years. The comet was re-observed at its return to perihelion in 1857, 1868, 1873, and 1879 ; but in 1884 it was looked for in vain. This comet was expected in February 1890, and several observers swept for it diligently, but to no purpose. Are we, therefore, to regard this as another lost member of our system ? Has the comet of Brorsen, like that of Biela, suffered dispersion in such degree as to be no longer within the reach of our powerful telescopes? Should negative results again attend observers in 1895, when the comet ought to return, there will be no reason to doubt its actual disappearance. It may be mentioned that owing to planetary perturbations, the period of this body has rapidly become shorter since 1846. It was then 2034 days, but in 1879 was reduced to 1995 days.

Faye's Comet.—First seen at the Paris Observatory on the night of Nov. 22, 1843, when it was near the star Bellatrix in Orion. The observations clearly proved the comet to be moving in an elliptical path, and Dr. Goldschmidt of Göttingen determined its period as $7\frac{1}{2}$ years. It was re-observed in 1851, and also during each of its five subsequent returns, the last of which occurred in August 1888. The orbit of this body approaches nearer to the circular form than that of any other known comet, except Tempel's of 1867. Its perihelion distance is considerable, for it never comes within the orbit of Mars. Prof. Möller, of Lund, has investigated the path with all the critical acumen of a profound mathematician, and, chiefly owing to his labours, it is now regarded as one of the best known members of our system.

D'Arrest's Comet.—Discovered at the Leipsic Observatory on June 27, 1851. M. Villarceau discussed the orbit, and announced it as an elliptic one with a periodic time of about $6\frac{1}{2}$ years. The comet was redetected at its return in 1857–8, 1870, 1877, and 1890. It is a very faint object.

Pons-Winnecke's Comet.—Discovered at Bonn on March 8, 1858, and on the elements being computed they were found nearly coincident with those of Pons's Comet, 1819 III. Encke had assigned a period of 5·62 years for the latter, but it managed to escape observation during the six returns that occurred in the 39 years between 1819 and 1858. Its identity was fully established in 1869, when it was again observed.

Tuttle's Comet.—A faint, diffused comet was discovered in the northern part of Hercules by H. P. Tuttle, of Cambridge, U.S.A., on Jan. 4, 1858. Its elements on being calculated were found by Pape to be similar to those of a comet discovered by Mechain on Jan. 9, 1790, and an elliptic orbit with a period of 13·66 years was derived from the new observations. On the assumption that the two bodies were one and the same there must have occurred four unobserved returns to perihelion between 1790 and 1858. The year 1871 was awaited in settlement of the question. When it came the comet returned, and the predictions received exact verification. Thus the comets of Mechain and Tuttle were placed in the inseverable bonds of identity.

Of the other periodical comets it will be unnecessary to give details. Some of them are still without full corroboration, only one return to perihelion having been observed. The reappearance of Pons's Comet (1812) in 1883–4, and of Olbers's Comet (1815) in 1887 furnished two excellent examples of well-determined comets belonging to the same class as that of Halley. Tempel's discoveries in 1867, 1869, and 1873 afforded some interesting additions to the family of short-period comets, and the list of these is continually extending owing to the assiduity of observers, though the lost comets of Biela and Brorsen will have to be removed from it. Peters's Comet of 1846 is also doubtful, as it escaped rediscovery in about 1859, 1872, and 1884 ; but this object may yet be captured at one of its succeeding apparitions. These bodies often evade redetection when their periods and paths are not accurately known. This has been fully exemplified in the case of the comets of Pons-Winnecke and Tuttle, which were unseen at several consecutive returns. It has been supposed, and not without reason, that the periodical comets are in process of wearing away. They apparently grow fainter at each return. Halley's Comet in 1835 was only moderately bright, whereas in ancient times its appearance was magnificent.

Grouping of Periodical Comets.—It is a curious circumstance that these bodies are assorted into groups having their aphelia near the orbits of major planets. The short-period comets comprised within the orbits of Encke's (3·29 years) and

R

Fig. 50.

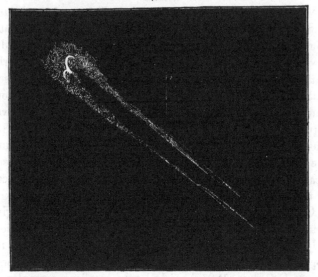

Pons's Comet (1812).   Telescopic view.   1884, January 6, 5ʰ 50ᵐ.
(10-inch reflector, power 60.   W. F. Denning.)

Fig. 51.

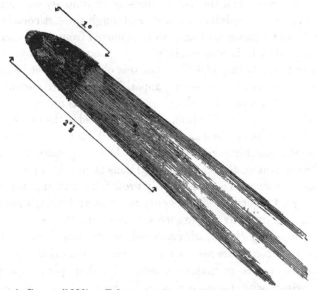

Pons's Comet (1812).   Telescopic view.   1884, January 21, 13ʰ 18ᵐ.
(5-inch refractor; comet-eyepiece, field 1¼°.   E. E. Barnard.)

Denning's (8·69 years) have aphelia in the region of the path of Jupiter, hence they are occasionally referred to as Jovian comets. The next group is represented by the comets of Peters and Tuttle, with aphelia near Saturn. The third group includes the comets of Tempel and Stephan, with aphelia just outside the orbit of Uranus. The fourth group is shown by the comets of Halley, Pons, and Olbers, with three others less certainly ascertained, with aphelia exterior to Neptune. There are unmistakable indications of other groups far outside the known boundaries of the solar system, but these are not so well defined. This clustering of cometary orbits has been ascribed to the attractive influences of the large superior planets, which are so capable of disturbing the paths of comets passing near that the orbits become transformed, and the aphelia henceforth lie near the points of extreme perturbation. This has been called the " capture theory;" and there is also an " ejection theory," which supposes the periodical comets to have had their birth in planetary ejections.

M. Hoek of Utrecht has found cases in which the orbits of two or more comets exhibit a common point of intersection in distant space, and infers their derivation from the same origin.

Further Observations required.—One of the chief and essential features in cometary work is the accurate determination of positions. But this entails the possession of expensive instruments, and a knowledge which amateurs have not always acquired. This department of labour can well be left to the trained hands at large observatories, where, fortunately, it meets with every attention. Ordinary observers will merely require to know the approximate place, and this is to be found by estimating the difference in R.A. and Dec. between a comet and a known star. The position of the latter may be found in a good catalogue and corrected for precession; then, allowing for the observed differences, the comet's place may be assigned to within very small limits of error. A low power, embracing a field of 1° or more, is best adapted for these observations, as it is more likely to include a catalogued star, and will exhibit the comet, especially if a large one, to the best effect.

The announcement of a new comet is always read with avidity by amateurs, and their first desire is to see it for themselves. This they may readily do by marking its place on a star-map or globe, and noting its relative place amongst the stars near. The telescope should then be directed towards the point indicated, and if the comet is not presented in the field, the instrument should be moved a little so that the surrounding region may be examined. If failure still attends the effort, the observer should point the telescope a few degrees E. or W. of the suspected point, and then carefully sweep over the place of the comet. It will then be picked up, unless it is too faint for his aperture. The first announcement of a comet generally gives the position at discovery, and the daily rate and direction of motion. The latter must of course be allowed for when the search is instituted.

The physical aspects of comets are as diversified as they are variable. No two comets are exactly alike, nor does the same comet exhibit a permanency of detail. Of course, when these objects are enormously distant, and barely visible, many of them appear to present similar characteristics ; but under the closer and more expanded views obtainable near perihelion the resemblance vanishes, and every comet is seen to possess features peculiar to itself. To trace these features, and to record them by delineation and description, forms one of the most interesting branches in which amateurs may engage. Much has been learnt of previous comets by successively noting their transitions of form and brightness, and the same scrupulous attention should be given to future comets.

The tails of comets are not always turned away from the Sun. Indeed, the contrary effect is sometimes produced. Occasionally there is a duple tail, the largest branch of which follows the normal direction, while the other is turned towards the Sun. Forms of this character require close watching from night to night. Is the sunward tail developed suddenly? and has it a fairly durable existence? Instances of singular curvature should also be noted. The tails are seldom perfectly straight, especially those attached to naked-eye comets, and decided changes affect their visible outlines at very short intervals. In large comets the space over which the tail extends should be sketched upon a star-map on successive evenings; its changes of position and curve will then be manifested by comparisons, and its increasing or decreasing length will also be apparent. Dark rifts, like shadows, often run lengthwise through the tail, and occasion a fan-like appearance analogous to that which distinguished the great comet of 1744 and gave it a sextuple tail.

The light of comets sometimes fluctuates in a very extraordinary manner, and too rapidly and irregularly to be consistent with theory. In this respect, Pons's Comet, at its last return in 1883-4 presented an eccentric behaviour. Bigourdan found that during the nineteen days from Sept. 5 to 24, 1883, the increase in the comet's brilliancy exceeded by thirty or forty times that resulting from reflected light alone ! This increase appears to have been due to a sudden outburst on Sept. 22, which occurred some time within the four hours preceding midnight. Dr. Müller, of Potsdam, witnessed a further outburst on Jan. 1, 1884, within $1\frac{3}{4}$ hour ; and the extent of this was accurately determined by means of a photometer. He found an augmentation of seven tenths of a magnitude in the brightness of the comet, and an equally sudden fall to its previous lustre. While these fluctuations were in progress, he noticed variations in the shape of the nucleus not less remarkable than its variations in light. Those who observe future comets will do well, therefore, to be on the alert for similar phenomena. The apparent brightness of the nuclei and alterations of shape or size should be recorded on every night when observations are feasible.

As a comet approaches the Sun its material apparently contracts, while with increasing distance from that luminary it expands. Usually the nucleus is extremely small and bright, and it often looks like a star shining through nebulosity. High powers must therefore be applied in its examination. Jets, aigrettes, luminous sectors, and other appendages are often involved with the nucleus and outlying coma, and they form a complicated structure well deserving further study. A good deal of mystery still surrounds these appearances ; their curious forms and vagaries have yet to be explained.

Stars are frequently observed through the head of a comet, which apparently, however, exercises no influence in dimming their lustre. But the stars are commonly seen behind the envelopes or comæ, and very rarely through the nucleus. Nothing is better calculated to exhibit the transparent and tenuous character of comets than observations of this kind, and observers should seek for further opportunities of making them. If the motion of a comet is obviously carrying it in the direction of one of the stars in the field, the observer may determine for himself the approximate time of conjunction by noting the distance between the star and comet and allowing for the motion of the latter. He will then know when to come to his telescope and witness the phenomenon. Should it appear probable that the comet's nucleus will pass over the star, he should commence his watch some time before it occurs ; he may then make comparisons before the star is involved in the outlying nebulosity, and trace the whole event from beginning to end. Any changes in the light or aspect of either star or comet would then be manifested. The comet of 1847 is said to have passed centrally over a 5th mag. star, but the latter was unaffected. Encke's Comet on one occasion interposed itself directly over one of a pair of 10th mag. stars, but their relatively equal brilliancy suffered no change. Encke's Comet, however, has no stellar nucleus. The latter feature is so bright and compact as displayed in many other comets, that its transit over a small star must have some effect either in obliterating it altogether, or in detracting from its lustre.

Visible evidences of rotation seem to have been suspected in certain comets, but this has never been substantiated on sufficient grounds. The circumstance is one, however, which should be remembered. During a series of observations the observer who notes the details of structure with particular regard to position may discover similar traces, and possibly learn something of the cause. The nucleus of a bright comet should always be examined with a moderately high power, so that any variations or peculiarities of form may be detected.

Nomenclature of Comets.—It must be confessed that no perfectly satisfactory method has yet been devised as regards the naming of comets. The plan of affixing Roman numbers progressively for each year, according to date of perihelion passage, answers pretty well, though a little confusion is sometimes caused by prematurely affixing the number, especially when two comets are discovered successively, the first of which is a long time before perihelion, and the second considerably after it. Until a comet can be safely assigned its catalogue place, it is preferable to refer to it by the name of the discoverer and date of discovery. This is more distinctive than the common method of lettering comets according to the epochs of their detection. As to periodical comets, it is not difficult to find some inconsistencies in their names. In the case of Halley's Comet (1682, discovered by Flamsteed) and Encke's Comet (1819 I., discovered by Pons), it was most fitting that they should be known to posterity by the names of the two able computers whose investigations first revealed to us comets of long and short period respectively. Under ordinary circumstances the name of the discoverer is applied to a comet as a means of convenient reference, and perhaps as a suitable recognition of the patient labours of the man who first announced it to the world. The plan seems to have been to name comets after those fortunate persons who sighted them at the particular apparition during which periodicity was determined. Thus Tuttle's Comet (1858 I.) had been seen as long before as 1790 by Mechain, and Biela's (1826 I.) was previously observed in 1772 by Montaigne, and in 1805 by Pons. It is, however, strange that a comet found by Pons in 1819 (III. of that year), and which Encke showed to be revolving in an ellipse with a periodic time of $5\frac{1}{2}$ years, should be called after Winnecke, who rediscovered it in 1858. To Pons the real priority belongs, though Winnecke deserves much praise for redetecting and identifying this body after it had effected six unobserved returns to perihelion. It is also curious to find that the comet of short period discovered by Swift in 1880 is called " Tempel's 3rd Comet" in Galle's catalogue (1885), from the fact that Tempel found it at a previous return (1869), when, however, its period was not ascertained. There is little doubt that the title justly belongs to Swift. Tempel himself called it " Swift's Comet." One plan should in fairness to observers be consistently adhered to. If comets are to be called after their original discoverers, then Biela's Comet should be known as " Montaigne's," Tuttle's as " Mechain's," &c.

Curiosities of Comets.—The comet of 1729, which was hardly visible to the naked eye, has far the greatest perihelion distance (4·0435) of any comet known. Barnard's Comet (1885 II.) comes next with a perihelion distance of 2·5068.

Pons's Comet at its return in 1883–4 remained visible for nine months. When last seen, on June 2, 1884, it was 470 millions of miles from the Earth, and more remote in the depths of space than any other observed comet since that of 1729. Barnard's Comet (1889 I.), though never visible to the naked eye, was followed from Sept. 2, 1888, to Aug. 18, 1890. Its distance from the Sun was then 6·25 (Earth's distance = 1), or about 580 millions of miles, which is greater than that of many of the short-period comets at aphelia. The most prolonged visibility of any previous comet was that of 1811 I. (510 days). But this comet of Barnard has been retained in view 715 days.

The great comet of 1882 was watched right up to the Sun's limb by Messrs. Finlay and Elkin at the Cape of Good Hope on Sept. 16, 1882. The comet was actually seen to disappear at the margin, and not a vestige of it could be traced during its transit across the solar disk. The nucleus of the comet was 4″ in diameter, and before transit it looked as bright as a part of the Sun's surface; but it was quite invisible when projected on the disk. The alleged observations by Pastorff and Stark, which were construed into visible transits of comets, are therefore thoroughly disproved, and will require another interpretation.

At the time of the total solar eclipse of May 17, 1882, a bright comet was observed near the Sun. It was a striking object visible to the naked eye. In the photographs which were taken of the eclipse the comet is well shown, but this body escaped subsequent observation, so that its orbit could not be determined.

Naked-eye Comets.—Arago mentions that twelve comets were visible to the naked eye during the period from 1800 to

1853, but there appear to have been certainly thirty comets fulfilling this condition, and I believe a careful search amongst cometary records would further augment the number. During the ten years from 1880 to 1889 inclusive there were no less than sixteen comets perceptible to unaided vision, and a considerable proportion of these were fine comets. It is very rarely that two naked-eye comets are to be seen at the same epoch, as in August 1881 and at the end of April 1886.

Comet-seeking.—For a long time after the invention of the telescope comet-seeking does not appear to have been undertaken in a methodical way, and to have formed the habitual work of certain observers. But the expected return of Halley's Comet in 1759 roused observers to take the initiative in a branch of practical research which in after years was destined to prove remarkably productive. Messier, Palitzch, and others began a system of sweeping the heavens for the predicted comet ; and it had a successful issue, for Palitzch, who did not relax his labours even on Christmas day, alighted upon the coveted prize on Dec. 25, 1758. Since that time a regular search after comets has been maintained. Messier pursued it with indomitable energy through a long period of years, and achieved many successes. It is said of him that on one occasion he was anticipated in a discovery by Montaigne, and he appears to have deplored the loss of the comet more than the loss of his wife, who was lying dead at the time. A friend visited him, and spoke a few words of sympathy in reference to his bereavement, but Messier, in despair about the comet, exclaimed : " I had discovered twelve—alas! that I should be robbed of the thirteenth by Montaigne ! " and his eyes filled with tears. Recollecting himself, and appreciating the loss he had sustained in his wife, he added, " Ah, this poor woman ! " Messier encountered some serious obstacles to his favourite pursuit. Breen, in his ' Planetary Worlds,' mentions that Messier, while walking in President Saron's garden, fell into an ice-house, and was disabled for a time. Later on " the revolution deprived him of his little income and every evening he was wont to repair to the house of Lalande to replenish the supply of oil for his midnight lamp. The political storm necessitated his removal to another neighbourhood, where he no longer heard the clocks of forty-two churches sounding the hours during his night-watchings." Messier discovered all his comets with a small 2-foot telescope of 2½ inches aperture magnifying 5 times and with a field of 4°.

Dr. Olbers, of Bremen, was another diligent student in this field. He did not effect many discoveries, but, from an upper apartment of his house, he observed nearly all the comets which appeared during half a century.

During the first twenty-seven years of the present century, Pons discovered the majority of the comets that were seen. He was a door-keeper at the Observatory at Marseilles, and owing to the teaching and encouragement he received from Thulis, the director, he achieved phenomenal success as a comet-hunter.

Discoveries of comets have rarely been effected in England. This is chiefly to be assigned to two circumstances. First, because the labour involved in seeking for these bodies has never perhaps been pursued to an equal degree and with the same tenacity as it formerly was in France, and as it has recently been in the United States ; and second, because the cloud-laden skies of England oppose the successful prosecution of a research in which a clear atmosphere is eminently desirable.

Though comet-seeking does not always produce new discoveries, it is certainly entertaining to those engaged in it; for one of the most agreeable diversions of telescopic work is to scan the firmament with a large-field comet-eyepiece, which exhibits the most pleasing views of star-groups, coloured stars, nebulæ, and telescopic meteors.

The operation of sweeping for comets is attractive from other aspects, though it undoubtedly needs close application, patience, and much caution. The possibility of seeing a comet in the field at any time proves a constant source of allurement to the observer, and sustains his enthusiasm. The glimpsing of a nebulous object, and the expectation (before it has been identified) that it *may* prove a comet, induces a little excitement which pleasantly relieves the monotony that might otherwise be attached to a sedulous research of this nature ; and it is one in which amateurs may suitably engage with a fair prospect of success. Instruments of great power, refinement, and expense are not required. It is rather a work calling for the exercise of patience and acute perception, and for that tireless servitude which those only who have an inborn love for it can maintain.

English Weather and Comet-seeking.—Only two new comets having been discovered in England during the last forty years some people regard our climate as in a great measure responsible for this. But the opinion seems to be erroneous. The lack of discoveries has arisen from want of effort as much as from want of opportunity. The best weather for comet-seeking is when the atmosphere is very transparent, and the stars are lucid and sparkling. Haze, fog, or cloud of any kind offers a serious hindrance. A thoroughly good night for planetary work is not usually good for cometary observation, because sharp definition is not so requisite as a very clear sky. A little fog or thin cloud, which will often improve planetary images, utterly obliterates a small telescopic comet. The air is sometimes very pure and dark after storms, and the stars remarkably bright ; it is then that the best opportunities are afforded for comet-hunting. Any systematic and regular work like this may be pursued in this country with every prospect of success by an observer who will persevere in it. From some statistics printed in the 'Science Observer,' Boston, it appears that during the seven months from May to November, 1882, Lewis Swift was comet-seeking during 300 hours. I have no English results of the same kind, but my meteoric observations will supply a means of comparison. From June to November, 1887 (six months), I was observing during 217 hours, and for a nearly similar period during the last half of 1877, though in each year work was only attempted with the Moon absent. My result for 1887 averages 36 hours per month, which is little less than the average derived from the comet-seeking records above quoted. It is therefore fair to suppose that as much may be done here as in some regions of the United States. Mr. W. R. Brooks wrote me in 1889, saying : " We have much cloudy weather in this part of America. While in other portions of the country clear weather abounds, it is not so in this section, where much

of my work has been done. This is a most fertile section—
the beautiful lake region of N.Y.,—but it is for this reason a
cloudy belt. It is far different in Colorado and California.
In the latter place, at the Lick Observatory, I hear they have
300 clear nights in the year—a paradise for the astronomical
observer! My former site, the Red House Observatory, Phelps,
N.Y., is only six miles from Geneva, and hence in the same
cloudy region."   Prof. Swift also referred to the subject of
weather in a letter to me dated July 30, 1889, where he says:
" I arrived home, after a five weeks' visit to the Lick Observatory, on March 1, and have not had half a dozen first-class
nights since—not in thirty years have I seen such prolonged
rainy and cloudy weather."   Now Mr. Brooks has discovered
13 comets in 7 years, Prof. Swift has found 8 comets (1862–
1890), and in addition to these has detected more than 700
new nebulæ, all of the latter since 1883. From this it appears conclusively that if such extensive and valuable results
can be obtained, notwithstanding frequently bad weather, then
English observers may prove equally successful, the important factor being that similar energy and ability direct their
labours.

Aperture and Power required.—Opinions are divided as to
the most suitable aperture and power for this work. Any
telescope of from 4- to 10-inches aperture may be employed
in it. A low power (30 to 50) and large field (50′ to 90′)
eyepiece are imperative ; and the instrument, to be really
effective, should be mounted to facilitate sweeping either in
a vertical or horizontal direction. A reflector on an alt-
azimuth stand is a most convenient form for vertical sweeps.
The defining-capacity of the telescope need not necessarily be
perfect to be thoroughly serviceable, the purpose being to
distinguish faint nebulous bodies, and not details of form. Far
more will depend upon the observer's aptitude and persistency
than upon his instrumental means, which ought to be regarded
as a mere adjunct to his powers and not a controlling influence
in success, for the latter lies in himself. Very large instruments are not often used, because of their necessarily restricted
fields. Moreover, a small instrument, apart from its advantage
in this respect, is worked with greater facility and expedition.

This is important, especially when the observer is to examine the region in the immediate neighbourhood of the Sun. He has then a very brief interval for the attainment of his purpose, and a small telescope must be used on account of its large field, its ready manipulation, and its general effectiveness on objects at low altitudes. The case is somewhat different when the search is to be conducted in regions far removed from the Sun's place ; for here the comets are in general faint, and there is time for the work to 'be deliberately and critically performed. Large instruments are to be recommended for these districts as capable of revealing fainter objects, though they are troublesome in several respects. They show large numbers of nebulæ, especially if the observer is exploring the region of Virgo, Coma Berenices, or Ursa Major ; and he will have great difficulty in identifying them and in feeling his way with certainty. These complications are inseparable from the work, and, though chiefly affecting large apertures, should not always be shunned ; for a telescope capable of displaying very faint nebulæ is also capable of showing faint comets. Many comets have eluded discovery by the inadequate reach of the instruments in the hands of comet-seekers ; and the statement recently made that there are only about one hundred nebulæ liable to be mistaken for comets is not accurate, because comets in certain positions are of the last degree of faintness, and there is no identifying them from small nebulæ except by means of their motion.

Mr. Brooks says :—" Medium magnifying powers, with necessarily moderate-sized fields, are better than very low powers and large fields. While with the latter a large amount of the sky can be swept over in a given time, the work is not so well done, and a faint comet would be easily swept over and not seen. A small region, *thoroughly worked*, is far more likely to be successful. This gives a feeling of satisfaction with the work performed, even with negative results. In support of this I may remark that, during all the years I have engaged in comet-seeking, not a single comet has been discovered by another astronomer in a region of the heavens that I had just previously searched ; so that I have never had occasion to feel that I had swept over a comet and missed seeing it. Aside from the obvious requirement of good eyesight, capable of detecting exceedingly faint objects, a good telescope of at least moderate aperture, and a familiarity attained by experience with the large number of nebulæ resembling telescopic comets, the comet-seeker, to be successful, must possess in a high degree the qualities of patience, perseverance, energy, and enthusiasm. I have the highest admiration for the man or woman who discovers a comet, because I know of the hard and thorough work which the success implies."

Mr. Brooks's experience and success in this branch give weight to his suggestions, and there can be little doubt that his commendation of moderate powers is fully justified. I believe he usually sweeps with a power of 40 (field of 1° 20′) on the 10½-inch equatoreal of his observatory. Speaking for myself, I find powers of 32 (field 1¼°) and 40 (field 1°) perform very satisfactorily on my 10-inch With-Browning reflector, having frequently tried them on faint nebulæ and comets. Sometimes I employ a power of 60, field 50′; but for ordinary purposes this is too high. It is a good plan to sweep with a moderate power, say of 40, and to keep a higher magnifier at hand to examine any suspicious objects that may be picked up. With power 32 I often encounter forms, the real character of which is uncertain. In such cases I clamp the telescope and apply the power 60, which generally exhibits the objects as several minute stars grouped together, or possibly nebulæ, in which case I proceed to identify them. With lower magnifiers than 30 there must always be considerable danger of sweeping over faint comets. Some of these are only of the 10th, 11th, or 12th mag., and less than 1′ diameter, and must certainly elude detection unless adequate power is brought to bear upon them. Dr. Doberck mentioned in the L. A. S. Journal, vol. vi. p. 236, an instrument for comet-seeking, 3½ inches in aperture, power about 10, and field of 5°, which was bought in 1842 by the late Mr. Cooper at Markree. But though with such a telescope a very large portion of the firmament might be swept in one night, there would be serious disadvantages; for small faint comets would pass through the field unseen, and render the work abortive. The necessary conditions of the

case go far to support the view that moderate powers and
fields are best ; for a search, to be thorough and satisfactory,
must be done critically, and with a power capable of revealing
the smallest specimens of comets.

Annual Rate of Discovery.—Arranging cometary discoveries during the century from 1782 to 1881 into periods of
20 years, and comparing the annual average with that during
the last eight years, we get the following numbers :—

Period.	Comets found.	Annual average.
1782–1801	25	1·25
1802–1821	26	1·30
1822–1841	36	1·80
1842–1861	83	4·15
1862–1881	79	3·95
1882–1889	40	5·00

These discoveries seem to have been greatly accelerated
about the year 1845. The yearly average between 1842 and
1881 was about 4 ; but between 1882 and 1889 it increased
to 5, owing mainly to the diligence of Barnard and Brooks.

The months in which the largest number of cometary discoveries have been effected are July and August, the figures
since 1782 being—

Month.	Comets found.	Percentage.
January	22	7·6
February	20	6·9
March	18	6·2
April	25	8·7
May	17	5·9
June	21	7·3
July	34	11·8
August	38	13·2
September . . .	22	7·6
October	20	6·9
November . . .	26	9·0
December . . .	26	9·0

Of 289 comets discovered during the last 108 years,

123 belonged to the first six months, while no less than 166 belonged to the last half of the year.

Though comets are not confined to any special region of the heavens, there is no doubt that the vicinity of the Sun is the spot to which the comet-seeker should direct his chief attention. It is here where the majority of the discoveries have been made ; and theoretically this should be so, seeing that the Sun is the controlling influence of the cometary flights, and that his position must be regarded as a sort of focus of their convergence and divergence. Hence the most likely spots are over the western horizon after sunset and the eastern horizon before sunrise. The twilight and zodiacal light, together with the mist at low altitudes, are impediments which are inseparable from this work ; but they need not interfere to any serious extent if the observer is careful to make the best of his opportunities. But though special attention is recommended to the neighbourhood of the Sun, other regions should not be altogether neglected, for comets are occasionally found in nearly the opposite part of the heavens to the Sun's place, as, for example, Zona's Comet of November 1890. In order to save time, and to prevent troublesome references during the progress of sweeping, the brighter nebulæ should be marked upon a star-chart, so that, as they enter the field, they may be instantly identified.

Telescopic Comets vary in size to a considerable degree. In diameter they generally range from about $1'$ to $7'$, and are usually round, with a bright centre like the globular clusters Messier 2, 3, 13, 15, 49, and 92, as seen with a low power ; but occasionally they are faint diffused masses, like the planetary nebula near β Ursæ Majoris, M. 97, or the large nebula S. of ζ Cassiopeiæ, in the New General Catalogue, No. 185, R.A. 0^h 33^m, Dec. $47°$ $44'$ N. In brightness they range from being visible to the naked eye to objects of the last degree of faintness. They average some $2'$ or $3'$ diameter, but are sometimes less than $1'$; so that the power of the sweeper should be capable of readily showing an object of this size as it passes through the field. The observer should turn his instrument upon the small planetary nebula N.G.C. 1501, R.A. 3^h 57^m, Dec. $60°$ $37'$ N. in Camelopardus. It is about 1' diameter. He should also pick up N.G.C. 6654, R.A. 18^h 27^m, Dec. 73° 6' N., which is a star of about 12½ mag. involved in a pretty conspicuous nebulosity. Swift describes the latter as looking just like a comet. N.G.C. 6217, R.A. 16^h 38^m, Dec. 78° 25' N., is also a small nebulosity which might easily be overlooked with a low power. Let the observer examine the three objects named, and he will gather a good idea of a small telescopic comet, especially from N.G.C. 6654, which may be readily found, as it is in the same field as χ Draconis, and visible at any time of the year and night. N.G.C. 6643, R.A. 18^h 23^m, Dec. 74° 32' N., is near the latter, but it is a brighter object. The observer will find two tolerably plain nebulæ in the same field at about R.A. 6^h 52^m, Dec. 85° 56'; so that they are only 4° from the pole. They are N.G.C. 2276 and 2300. These objects ought not to elude detection in any instrument properly adapted for comet-seeking.

Ascertaining Positions.—No observer should be without the means of determining exact positions. A ring-micrometer and comprehensive star-catalogues are most important accessories of the amateur. When a suspicious object is found its precise position should be instantly measured; but if no micrometer is at hand, the observer should carefully note the place relatively to adjoining stars, and then, after a short interval, re-observe it for traces of motion. In these comparisons the low-power eyepiece should be exchanged for one of greater amplification, because this will render a slight motion more readily sensible. If the suspicious object proves to be a comet, the extent and direction of its daily motion should be computed from the change in the observed places, and the information telegraphed to the Royal Observatory, Greenwich. A statement should also be given as to the diameter and brightness of the object; we may then be satisfied that it will be readily picked up at some of the many stations where prompt attention is given to this class of observation. Amateurs who do not attempt to obtain exact positions are sometimes condemned for their negligence in this respect, and most unjustly so. By far the hardest part of the work falls to them, and professional astronomers ought to be indebted to amateurs for leaving to their care an important feature of these observations. If the latter are to undertake the labour of measuring as well as discovering comets, then there will be nothing left in this line for the elaborate instruments of observatories to do. Yet, while thus objecting to amateurs, with their generally incomplete and inefficient appliances, being expected to perform the work both of discovery and exact observation, it cannot be denied that there is a great necessity for them to have the means of measurement, and to utilize them during the first few observations, which are usually made before the comet has been seen elsewhere, and will therefore possess great value if precise.

Dr. Doberck's Hints.—Dr. Doberck has given some useful hints in connection with this subject:—" In order to be as sure as possible of ultimate success it is not enough to sweep with the instrument and watch any suspicious object for proper motion. It is better to procure a large map such as Argelander's, and, comparing the image seen in the comet-seeker with the map, to insert all the nebulous objects according as they are discovered. At the end of the watch they are then compared with the catalogues of nebulæ and clusters of stars. A general catalogue facilitates this, but is never quite sufficient, as there seems to be no limit to the number of objects in the sky, and more are constantly being catalogued. In the course of time an observer learns to remember the objects he has seen before in the seeker, and at last he need not consult the map at all. The subsequent observation of a newly-found comet is best made with the ring-micrometer if the telescope is not equatoreally mounted. In the latter case it should be made by aid of a steel-bar micrometer. As soon as three observations are available the first approximation to a parabolic orbit can generally be determined: the calculation of which is quite elementary, and would be enjoyed by many amateur astronomers who are fond of figures and would easily get used to Olbers's method. Only the three positions must not be so near each other as to lie on a great circle."

Prizes for Discoveries.—The Vienna Academy of Sciences formerly gave a gold medal to the discoverer of every new comet. These presentations were discontinued in about 1880 ; and Mr. H. H. Warner then offered a prize of $200 for every unexpected comet found in the United States or in Canada. This prize was continued in subsequent years, and the conditions were amended so as to include observers in Europe. Many of these prizes were gained by Barnard and Brooks ; but they have not been re-offered during the past year or two. Mr. Warner, however, contemplates renewing them. The Astronomical Society of the Pacific now awards a bronze medal to all such discoverers.

CHAPTER XV.

METEORS AND METEORIC OBSERVATIONS.

Ancient ideas concerning Meteors.—Meteoric Apparitions.—Radiation of Meteors.—Identity of Meteors and Comets.—Aerolites.—Fireballs.—Differences of Motion.—Nomenclature of Meteor-Systems.—Meteor-Storms.—Telescopic Meteors.—Meteor-Showers.—Varieties of Meteors.—Heights.—Meteoric Observations.

> " As oft along the still and pure serene
>   At nightfall, glides a sudden trail of fire,
>   Attracting with involuntary heed
>   The eye to follow it, erewhile it rest;
>   And seems some star that shifted place in heaven."
>
>   DANTE.

No one can contemplate the firmament for long on a clear moonless night without noticing one or more of those luminous objects called shooting-stars. They are particularly numerous in the autumnal months, and will sometimes attract special attention either by their frequency of apparition or by their excessive brilliancy in individual cases. For many ages little was known of these bodies, though some of the ancient philosophers appear to have formed correct ideas as to their astronomical nature. Humboldt says that Diogenes of Apollonia, who probably belonged to the period intermediate between Anaxagoras and Democritus, expressed the opinion that, " together with the visible stars, there are invisible ones which are therefore without names. These sometimes fall upon the Earth and are extinguished, as took place with the star of stone which fell at Ægos Potamoi." Plutarch, in the ' Life of Lysander,' remarks :—" Falling stars are not emanations or rejected portions thrown off from the ethereal fire, which when they come into our atmosphere are extinguished after being kindled : they are, rather, celestial bodies which, having once had an impetus of revolution, fall, or are cast down to the Earth, and are precipitated, not only on inhabited countries, but also, and in greater numbers, beyond these into the great sea, so that they remain concealed."

In later times, however, opinions became less rational. Falling stars were considered to be of a purely terrestrial nature, and originated by exhalations in the upper regions of the air. Shakespeare expressed the popular belief when he wrote :—

> " I shall fall
> Like a bright exhalation in the evening,
> And no man see me more."

Another theory, attributed to Laplace, Arago, and others, was that meteors were ejections from lunar volcanoes. But these explanations were not altogether satisfactory in their application. The truth is, that men had commenced to theorize before they had begun to observe and accumulate facts. They had learnt little or nothing as to the numbers, directions, and appearances of meteors, and therefore possessed no materials on which to found any plausible hypothesis to account for them.

Meteoric Apparitions.—The occasional apparition of brilliant detonating fireballs, the occurrence of remarkable star-showers, the precipitation upon the Earth's surface of stony masses, were facts which could be verified from many independent sources, and they set men thinking how to account for the strange and startling freaks of nature as exhibited in such phenomena. But though records existed of exceptionally large meteors and of meteor-showers, the descriptions were imperfect and failed in the most important details. The observers were usually unprepared for witnessing such events, and gave exaggerated and inaccurate accounts of what they had seen. The vivid brightness of a fireball (overpowering the lustre of the stars, and even vieing with the Moon in splendour), the flaming train left in its wake (curling itself up into grotesque shapes, as it drifted and died away), the form of the nucleus with its jets and sparks, and the final explosion, with the reverberations it caused, were all alluded to by the enthusiastic observer ; but it was only in rare cases that the more valuable features were placed on record. The *direction* and *duration* of the meteor's flight amongst the stars were facts of greater significance than the mere visible aspect of the object ; but they were seldom regarded. Hence the early observations proved of little weight in inducing just conceptions as to the phenomena of meteors.

There is, perhaps, no celestial event which can compare, as regards its striking aspect and interesting features, with that of a meteoric display of the most brilliant kind. A large comet, a total solar eclipse, a bright display of aurora, have each their attractive and imposing forms ; but the effect produced is hardly equal to that during the Earth's *rencontre* with a dense meteor-swarm. The firmament becomes alive with shooting-stars of every magnitude ; their incessant flights are directed to every point of the compass for several hours ; and the scene is so animated, and one of such peculiarly impressive and novel character, that it can never be forgotten by those who have been among its fortunate spectators.

Radiation of Meteors.—Heis, in Germany, was the pioneer in this branch of practical astronomy. About half a century ago he began systematic observations, and gathered many useful data. Schmidt, at Bonn and Athens, followed his example ; and in England Prof. Alexander Herschel and Mr. R. P. Greg devoted themselves to the subject with highly successful results. Their collective labours revealed a large number of well-defined systems of meteors, and enabled them to publish tables of the radiant-points. The investigations were more precise than formerly, and conducted on methods ensuring more accurate and plentiful materials. The radiation of meteors from fixed points in the sky had been observed before in regard to the great display which occurred in November 1833 ; but the meteors that fell on ordinary nights were regarded as sporadic, until Heis and his immediate successors showed they were reducible to an orderly arrangement and that every one of them had its radiant-point and its origin in a definite meteor-stream. The apparently divergent flights from a common centre are simply due to the effects of perspective on bodies really moving in parallel directions and collected into groups more or less scattered.

Identity of Meteors and Comets.—The mystery concerning

Fig. 52.

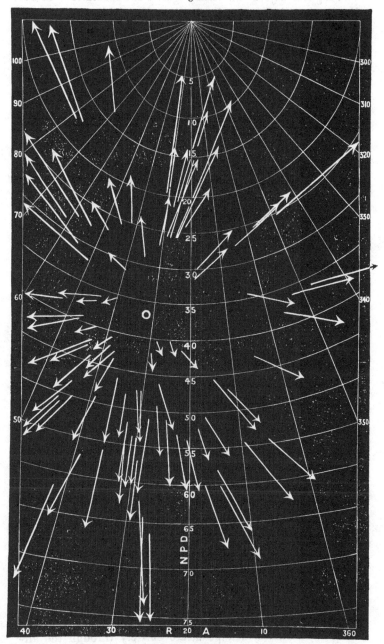

Radiation of Meteors

(Shower of early Perseids from 32°+53°, July 28–Aug. 1, 1878.)

these fugitive objects and their vagaries of appearance was not always to remain concealed. Denison Olmsted had, in his work on 'The Mechanism of the Heavens,' published in 1850, stated that the constitution of the body to which the meteors of 1833 belonged bore " a strong analogy to comets." Reichenbach, in 1858, wrote a paper in which it was sought to prove that a comet is a swarm of meteorites. Prof. Kirkwood, in 1861, also concluded that " meteors and meteoric rings are the debris of ancient but now disintegrated comets, whose matter has become distributed around their orbits." But it remained for Schiaparelli, of Milan, in 1866, to demonstrate the identity of meteoric and cometary systems. Others had reasoned up to it, and observers had amassed many useful observations bearing on the subject; but absolute proof was wanting until Schiaparelli supplied it. He computed elements for a well-known shower of meteors occurring on August 10th, and found the orbit presented a very close resemblance to that of Comet III. 1862; and he detected a similar analogy between the November meteors and Comet I. 1866. The orbit of the April meteors was afterwards shown by Galle and Weiss to agree with the path of Comet I. 1861; and a meteor-shower occurring at the end of November was found to coincide with Biela's Comet. Facts like these could not be disproved. Comets were thenceforth known to be the parents—the derivative source—of meteors. Thus two important classes of objects became as one, the differences observed being merely those of aspect due to the variable conditions under which they were presented. The great meteor-shower of November was found to be the dispersed materials of Tempel's Comet of 1866 seen in detail and from a near standpoint. Every meteoric display was known to be the visible effects of the collision of the Earth with a comet or with the great stream of planetary fragments describing a cometary orbit.

Aerolites.—Meteors enter our atmosphere with such great velocity that the friction induced by their impact is sufficient to destroy them by combustion. They rarely approach the Earth's surface within 15 miles. Occasionally, however, a slow-moving meteor of large size, and formed of a very compact substance, will penetrate entirely through the air-

Fig. 53.

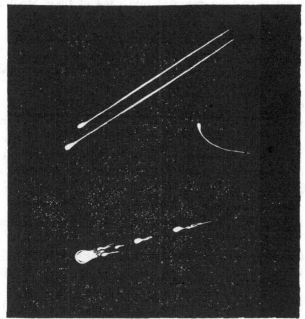

Drawn by W. F. Denning

1. Double meteor, Dec. 29, 1886.   2. Curved meteor, Dec. 25, 1886.
3. Fireball, Sept. 7, 1888.

Fig. 54.   Fig. 55.

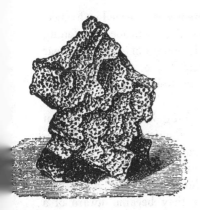   

Meteorite found in Chili in 1866.   Meteorite which fell at Orgueil in 1864.

strata and fall upon the Earth's surface. Many instances of
the kind have been recorded, and a few of these are quoted
below:—

1478 B.C. The Parian chronicle records that an aerolite or
thunder-stone fell in the island of Crete. This
appears to be the earliest stone-fall described in
history.

654 B.C. A shower of stones descended near Rome.

465 B.C. A stone, surrounded with fire, fell in Thrace.
This stone is referred to by several ancient writers.
It was termed the "Mother of the Gods," and is
said to have fallen at the feet of the poet Pindar.

52 B.C. A shower of iron descended at Lucania, in the
time of Crassus.

1492 A.D. A stone weighing 262 lb. fell at Ensisheim, in
Alsace.

1642. A stone of 4 lb. fell near Woodbridge, in Suffolk.

1795, Dec. 13. A stone of 56 lb. fell at Wold Cottage,
Thwing, Yorkshire.

1860, July 14. A shower of aerolites fell at Dhurmsala,
in India. A tremendous detonation attended their
descent, and the natives became greatly alarmed.
They supposed the stones to have been thrown by
some of their deities from the summit of the Himalayas, and many of them were preserved as objects
of religious veneration.

1864, May 14. A very large meteor was observed in France.
At Montauban and the neighbourhood deafening
explosions occurred, and showers of stones fell near
the villages of Orgueil and Nohic.

1876, April 20. A piece of iron weighing 7¾ lb. fell at
Rowton, Shropshire.

1881, March 14. A stone weighing 3 lb. 8¼ oz. fell at
Middlesborough, Yorkshire, on a part of the North-
Eastern Railway Company's branch line. The
descent of the aerolite was witnessed by an inspector
and three platelayers, who were working about fifty
yards distant. At first they became aware of a
whizzing or rushing noise in the air, immediately

followed by the sudden blow of a body striking the ground near. The hole, 11 inches deep, which the stone made was found directly after, and the stone was extracted.

Many other examples might be given, but the above will be sufficient for our purpose. Records of this nature were discredited in former times ; but more modern researches have long since placed their reality beyond all question. The fall of stones from the sky is no longer regarded as a mere legendary tale, but as one of the well-assured operations of nature.

Meteoric stones and irons have been classified according to the ingredients of their composition. Those in which iron is found in considerable amount are termed siderites, those containing an admixture of iron and stone, siderolites, and those consisting almost entirely of stone are known as aerolites. The siderite which fell in Shropshire on April 20, 1876, forms only the seventh recorded instance where a mass of meteoric iron has been actually seen to fall.

Fireballs.—The table on p. 268 gives the dates, heights, &c. of fifteen fireballs observed during the last quarter of a century.

Fireballs are sometimes detonating, though more often silent. The fireball of Nov. 23, 1877, gave a sound like salvoes of artillery, and doors and windows were shaken violently. At Chester the noise of its explosion was compared to loud but distant thunder. Lieut.-Col. Tupman says that "thunder, to be loud, must be within five miles ; hence it appears that the violence of the explosion must have been at least a hundred times greater than a peal of thunder, the intensity of sound-waves diminishing as the square of the distance." "The explosion of a 13-inch bomb-shell, consisting of some 200 lb. of iron, would not have produced a sound of one hundredth part of the intensity of the meteor-explosion." This fireball must therefore have been an object of considerable mass before its dissolution ; and it is fortunate that such bodies are usually destroyed by the effects of combustion before they reach the Earth's surface.

These phenomena exhibit many varieties of appearance.

Date of Apparition.	G.M.T.	Height. At Appearance.	Height. At Disappearance.	Real Length of Path.	Velocity.	Radiant-Point. R.A. Dec.	Authority.
	h m	miles.	miles.	miles.	miles.		
1865, April 29	12 42	52	37	75	20	73 +47	A. S. Herschel.
1868, Sept. 5	8 5	250	85	1200	28	14 − 2	G. von Niessl.
1869, Nov. 6	6 50	90	27	170	35	62 +37	A. S. Herschel.
1872, July 22	8 55	77	37	88	.....	246 −11	T. H. Waller.
1874, Aug. 10	11 53	77	33	105	17	325 −17	W. H. Wood.
1875, Sept. 3	9 55	75	40	35	27	311 +52	G. L. Tupman.
1875, Sept. 14	8 28	52	13	104	13	348 − 0	G. L. Tupman.
1876, Sept. 24	6 30	58	16	45	15	285 +35	A. S. Herschel.
1877, Nov. 23	8 25	95	14	135	17¼	62 +21	G. L. Tupman.
1878, June 7	9 53	65	37	160	19	247 −25	A. S. Herschel.
1879, Feb. 23	14 53	60	7	102	14½	310 +55	J. E. Clark.
1886, Nov. 17	7 18	96	21	123	17½	34 +19	W. F. Denning.
1887, May 8	8 22	70	14	110	18	191 − 5	W. F. Denning.
1888, Aug. 13	11 33	78	47	46	.....	43 +56	W. F. Denning.
1889, May 29	10 44	58	23	76	8½	216 7	D. Booth.

Fig. 56.

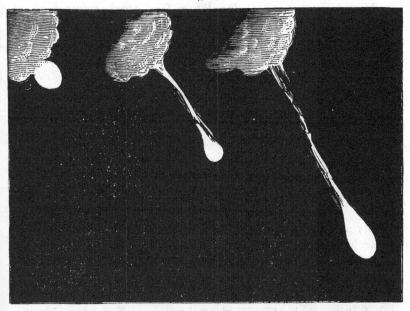

Fireball of Nov. 23, 1877, 8h 24m, emerging from behind a cloud.
(Drawn by J. Plant, Salford.)

Sometimes there is no visible explosion; the bright nucleus slowly dies out until reduced to a faint spark before final disappearance. Several outbursts of light are often noted; and a curious halting motion has been observed in regard to large slow-moving meteors. I have occasionally remarked a succession of four brilliant flashes given by individual fireballs. These flashes, though sometimes of startling intensity, are somewhat different to the transient vividness of lightning; they come more softly, and remind one forcibly of moonlight breaking suddenly from the clear intervals in passing clouds.

Fireballs differ vastly from shooting-stars in point of size; but their origin is identical. The August meteor-shower yields the smallest shooting-stars and the largest type of fireballs. The great display of meteors on Nov. 27, 1885, not only presented us with large and small members, but it also furnished us with a siderite or piece of iron, presumably from Biela's Comet. This fell at Mazapil, Mexico; and as considerable interest is attached to the case, I quote a part of the discoverer's statement:—

" It was at about 9 o'clock on the night of November 27th, when I went out to the corral to feed certain horses : suddenly I heard a loud sizzing noise, exactly as though something red-hot was being plunged into cold water; and almost instantly there followed a somewhat loud thud. At once the corral was covered with a phosphorescent light; while suspended in the air were small luminous sparks, as though from a rocket. . . . A number of people came running towards me; and when we had recovered from our fright we saw the light disappear, and bringing lanterns to look for the cause found a hole in the ground, and in it a ball of light. We retired to a distance, fearing it would explode and harm us. Looking up to the sky, we saw from time to time exhalations of stars, which soon went out without noise. We returned after a little, and found in the hole a hot stone which we could barely handle; this, on the next day, we saw looked like a piece of iron. All night it rained stars; but we saw none fall to the ground, as they all seemed to be extinguished while yet very high up."

This is the first observed instance in which a meteorite has actually reached the Earth's surface during the progress of a star-shower. If its identity with the meteors of Biela's Comet is admitted, then all classes of meteoric phenomena would appear to have a community of origin.

Differences of Motion.—Great differences are observed in the velocity of meteors. An observer may notice all varieties on the same night of observation. Some will move very slowly, others shoot quickly across the sky. These differences are occasioned by the astronomical conditions affecting the position of the meteor-orbit relatively to the motion of the Earth. Thus the meteors of Nov. 13 move with great velocity (44 miles per second), because they come directly from that part of the heavens towards which the Earth is moving; hence the orbital speed of the Earth (18½ miles per second) and meteors (26 miles per second) is combined in the observed effects. But in the case of the meteor-shower of Nov. 27 the motions are extremely slow (about 10 miles per second), as the Earth and the meteors are travelling nearly parallel in the same direction, and the latter have to overtake the Earth.

Nomenclature of Meteor-Systems.—It is customary to name the showers after the constellation from which the meteors appear to diverge. Thus the meteors of April 20 are called *Lyrids*, the radiant being in Lyra ; the meteors of August 10 are termed *Perseids*, the point of emanation being in Perseus. The two great streams of November are known as the *Leonids* (13th) and *Andromedes* (27th). Several showers are often visible in the same constellation ; and when it is desired to name these according to the above system, it is necessary to add the approximate star to distinguish them. Thus, in August there are showers of μ Perseids, ε Perseids, and α Perseids, in addition to the well-known Perseids of August 10.

Meteor-Storms.—On Nov. 12, 1799, Humboldt, at Cumana, in South America, saw " thousands of bolides and falling stars succeed each other during four hours." On Nov. 12, 1833, this shower recurred, and was witnessed with magnificent effect in America. One observer stated that between 4 and 6 A.M. (Nov. 13) about 1000 meteors per minute might have been counted ! Another display occurred on Nov. 13,

1866, and on this occasion 8485 meteors were enumerated by several observers at Greenwich. A different system gave us a brilliant exhibition on Nov. 27, 1872, when 33,000 meteors were counted by Denza and his assistants at Moncalieri, in Italy, between the hours of 5^h 50^m and 10^h 30^m P.M. A repetition of this phenomenon occurred on Nov. 27, 1885, when the same observers counted nearly 40,000 meteors between 6^h and 10^h P.M.

Telescopic Meteors.—Observers who are engaged in seeking for comets or studying variable stars employ low powers and large fields, and during the progress of their work notice a considerable number of small meteors. At some periods these bodies are more plentiful than at others, and appear in such rapid succession that the observer's attention is dis-

Fig. 57.

VERTICAL.

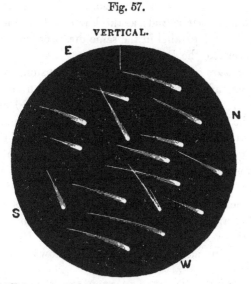

Flight of Telescopic Meteors seen by W. R. Brooks, Nov. 28, 1883.

tracted from the special work he is pursuing to watch them more narrowly and record their numbers. Schmidt saw 146 telescopic meteors during ten years. They ranged between the 7th and 11th mags. Winnecke in the year 1854 noticed 105 of these objects on thirty-two evenings of observation with a 3-inch finder, power 15, and field of 3°. I have also remarked many of these objects when using the comet-eyepieces of my 10-inch reflector *, and find they are apparently more numerous than the ordinary naked-eye meteors in the proportion of 22 to 1. It would be supposed from the great rapidity with which the latter shoot across the firmament that the smaller telescopic meteors are scarcely distinguishable by their motion, as they must dart through the field instantaneously and only be perceptible as lines of light. But this impression is altogether inconsistent with the appearances observed. They possess no such velocity, but usually move with extreme slowness, and not unfrequently the whole of the path is comprised within the same field of view. The eye is enabled to follow them as they leisurely traverse their courses, and to note peculiarities of aspect. Of course, there are considerable differences of speed observed, but as a rule the rate is decidedly slow and far less than that shown by naked-eye meteors. I believe that telescopic meteors are situated at great heights in the atmosphere, and that their diminutive size and slowness of movement are due to their remoteness. This conclusion will hardly be avoided by anyone who attentively studies the several classes of meteors in their various aspects. Unfortunately no attempt appears to have been hitherto made to determine the actual heights of telescopic meteors, owing to the difficulty of obtaining two reliable observations of the same object. The only way of securing such data would be for several observers to watch certain selected regions by prearrangement either with a low-power telescope or field-glass, and record the exact times and paths of the meteors seen. On a comparison of the results a good double observation of the same object might be found, in which case the real path could be readily computed.

Future observers should note the different forms of telescopic meteors. Safarik has divided them into four classes. viz. :—(1) Well-defined star-like objects of very small size ; (2) Large luminous bodies of some minutes of arc in diameter ; (3) Well-defined disks of a very perceptible diameter

* During the seven months from May to November 1890 I noted ninety-five telescopic meteors while engaged in comet-seeking.

brighter at the border than at the centre, which gives them
the aspect of hollow transparent shells ; and (4) faint diffused
masses of irregular shape, considerable size, and different
colours. He has seen hundreds of meteors of every magnitude from the 2nd down to the 12th pass through the field
of his 6½-inch reflector (ordinary power 32, field 54′). On
Aug. 30, 1880, 9^h to 15^h he observed between 50 and 100
telescopic meteors, and many others were seen on the following
night. Whenever a shower of these bodies, such as that witnessed by Brooks on Nov. 28, 1883, occurs, observers should
notice whether the objects participate in a common direction
of motion ; because, if so, the radiant-point will admit of determination. The horary rate of their apparition ought also to be
ascertained. Those who habitually search for comets should
invariably make a note of telescopic meteors, as such records
would aid inquiries into the relative frequency of these
phenomena.

Meteor Showers.—The following short list includes the
principal displays of the year :—

Name of Shower.	Duration.	Date of Max.	Radiant-Point.	Sun's Longitude.
			α　　　δ	
Quadrantids	Dec. 28–Jan. 4	Jan. 2	229·8+52·7	281·6
Lyrids	April 16–22	April 20	269·7+32·5	31·3
η Aquarids............	April 30–May 6	May 6	337·6– 2·1	46·3
δ Aquarids............	July 23–Aug. 25	July 28	339·4–11·6	125·6
Perseids	July 8–Aug. 22	Aug. 10	45·9+56·9	138·5
Orionids...............	Oct. 9–29	Oct. 18	92·1+15·5	205·9
Leonids	Nov. 9–17	Nov. 13	150·0+22·9	231·5
Andromedes	Nov. 25–30	Nov. 27	25·3+43·8	245·8
Geminids	Dec. 1–14	Dec. 10	108·1+32·6	259·5

Notes.

Quadrantids. Heis was the first to determine this radiant
accurately. It was subsequently observed by Masters and
Prof. Herschel (1863–4). The radiant is circumpolar in this
latitude, but low down during the greater part of the night,

hence the display is usually seen to the best advantage on the morning of Jan. 2.

Lyrids. Attention was first drawn to the April meteors by Herrick in the United States. Active displays occurred in 1863 and 1884.

η Aquarids. Further observations are urgently required of this stream. The radiant is only visible for a short time before sunrise. There is a considerable difference between my results and those secured by Lieut.-Col. Tupman, the discoverer of this system in 1870, whose observations place the radiant at $326\frac{1}{2} - 2\frac{1}{2}$ April 29–May 3. These May Aquarids are interesting from the fact that they present an orbital resemblance to Halley's Comet, which makes a near approach to the Earth on May 4, twelve days before reaching the descending node.

δ Aquarids. The meteoric epoch, July 26–30, was first pointed out by Quetelet many years ago. Biot also found, from the oldest Chinese observations, a general maximum between July 18 and 27 (Humboldt). Showers of Aquarids were remarked by Schmidt, Tupman (1870), and others; but it was not known until my observations in 1878 that the Aquarids formed the special display of the epoch, and that there were many early Perseids visible at the same time.

Perseids. Muschenbroeck, in his work on 'Natural Philosophy,' printed in 1762, mentions that he observed shooting-stars to be more numerous in August than in the other months of the year. Quetelet, in 1835, was, however, the first to attribute a definite maximum to the 9th–10th. This stream is remarkable for its extended duration, and for the obvious displacement which occurs from night to night in the place of its radiant. It furnishes an annual display of considerable strength, and is, perhaps, the best known system of all.

Orionids. Profs. Schmidt and Herschel were the first to discover the Orionids as the most brilliant display of the October period, and accurately determined its radiant in 1863-4-5. Herrick recorded a shower at $99° + 26°$, Oct. 20–26, 1839, and Zezioli in 1868 recorded many meteors which were ascribed to a radiant at $111° + 29°$; but there is no doubt that the Orionids were observed in both these cases, though the radiant was badly assigned.

The radiant of the Orionids shows no displacement like that of the Perseids.

Leonids. Observed from the earliest times. Humboldt and Bonpland saw it well on the night of November 11-12, 1799, and the phenomenon at its magnificent return on November 12, 1833, was ably discussed by Olmsted. It furnished a splendid shower in 1866, November 13, and many meteors were seen at the few subsequent returns. I observed fairly conspicuous showers of Leonids in 1879 and 1888. There is no doubt the meteors form a complete ellipse, for the earth encounters a few of them at every passage through the node. Grand displays may be expected at the end of this century.

Andromedes. Observed by Brandes, at Hamburg, Dec. 7, 1798. It also recurred in 1838; the very brilliant showers of November 27, 1872 and 1885, are still fresh in the memory. It is uncertain whether this group forms an unbroken stream; if so, the regions far removed from the parent comet must be extremely attenuated. Some of the meteors were seen in 1877 and 1879. The radiant is diffuse to the extent of 7° or 10°. Returns of the shower should be looked for in 1892 and 1898.

Geminids. Mr. Greg first called attention to the importance of this shower. It was well observed by Prof. Herschel in 1861-3-4, and some later years.

There are an enormous number of minor systems, but these are generally feeble, and interesting only to the regular observer of meteors. Many showers are so slightly manifested that they yield but one visible meteor in 6 or 7 hours, and on the same night of observation there are often as many as 50 or 60 different systems in operation. I gave a list of 918 radiant-points of showers observed at Bristol in the ' Monthly Notices,' May 1890, and other catalogues will be found in the ' British Association Reports ' for 1874 and 1878.

Varieties of Meteors.—The amateur who systematically watches for meteors will occasionally remark instances of anomalous character. I have sometimes observed meteors which are apparently very near, and move with enormous velocity. They are mere gleams of pale light, which have little analogy to ordinary shooting-stars, and suggest an electric origin, though I do not know whether the marvellous quickness with which they flash upon the eye is not to be held responsible for the impression of nearness. They are some-what rare, and one may watch through several entire nights without a single example, but as far as my memory serves I must have witnessed some scores of these meteoric flashes.

One of the most interesting class of meteors includes those which move so slowly that the eye is enabled to note the details of their appearance. Some of these objects are small when first seen, but enlarge considerably under the increasing temperature, and after a great slackening of speed (due to atmospheric resistance) their nuclei are finally spent in thick streams of luminous dust. On Dec. 28, 1888, I recorded a meteor which on its first apparition was tolerably bright,

Fig. 58.

Meteor of Dec. 28, 1888, 6ʰ 17ᵐ.

small, and compact. It moved slowly, and I had an excellent view of its passage. The nucleus quickly expanded, though with no increase of brilliancy. Towards the end it assumed a sensible disk, and at the last phase the mass spread or de-ployed itself into a wide stream of fine ashes and disappeared. The whole phenomenon was so curious, and observed with such distinctness, that I made the above sketch of it directly afterwards.

Heights of Meteors.—Usually the height of meteors at their first appearance is less than 90 miles, and at disappearance more than 40 miles. From a comparison of a large number of computations I derived the following average values :—

Beginning height . . . 76·4 miles (683 meteors)
End height . . . . . 50·8 „ (756 „

But if fireballs and the smaller shooting-stars are separated I find the usual heights at disappearance are :—fireballs, 30 miles; shooting-stars, 54 miles. Fireballs therefore approach much nearer to the Earth's surface before disruption than the ordinary falling stars.

A very slight acquaintance with trigonometry will enable anyone to compute the real path of a meteor if two or more observations, made at distant stations, are available for the purpose. The observed courses of the meteor should be marked upon a celestial globe, and extended backwards to the point where they mutually intersect; this will be the

Fig. 59.

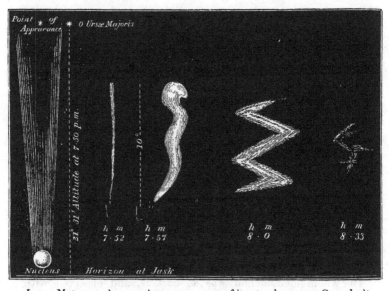

Large Meteor, and successive appearances of its streak, seen at Cape Jask, in the Persian Gulf, on June 8, 1883, 7ʰ 51ᵐ to 8ʰ 33ᵐ.

radiant-point. The globe having been set for the time and latitude, the apparent tracks should also be prolonged in a forward direction until they meet the horizon, this will indicate the *Earth-points,* or azimuths of the place where the meteor would have been precipitated on the Earth had it been enabled to continue its flight so far. The azimuths and altitudes of the beginning and end of the path, and the azimuths of the Earth-point should then be read off, and by means of a reliable map and a protractor their points of intersection over the Earth's surface may be readily found by lines drawn from the two places of observation. From the spot where the Earth-points intersect a straight line should also be drawn in the direction of the radiant, and it is along this line the meteor's motion was directed. The coordinates of the observed points of appearance and disappearance of the meteor, at the two stations, would intersect this line at identical points were the observations perfectly accurate, but this is rarely the case. The distance between the observer's station and the places over which the meteor began and ended is easily derived from the map, and the height of the object may be found by adding the logarithm of the distance to the log. of the tangent of the altitude. Thus, if the end of a meteor is witnessed from London in azimuth 130° W. of S. (alt. 25°), and from Bristol in azimuth 216° W. of S. (alt. 30°) the place of intersection on the map will be at Warwick, so that the meteor must have disappeared when vertically over this city. London is distant from Warwick about 86 miles, and from Bristol 70 miles, and the resulting height of the meteor is:—

	London.			Bristol.
86 log.	1·93450		70 log.	1·84510
25° tan	9·66867		30° tan	9·76144
	1·60317 = 40·1			1·60654 = 40·4

so that the observations accord very closely in fixing the height at a little exceeding 40 miles at disappearance, but a slight correction is necessary to allow for the Earth's curvature. There are other methods of computing the heights, one of which is explained by Prof. A. S. Herschel in a paper entitled " Height of a Meteor " (' Monthly Notices,' vol. xxv. p. 251).

Meteoric Observations.—A large number of meteor-showers still await discovery, and there are features even in connection with the best known streams which remain to be elucidated. Such doubts as now exist are only to be cleared away by assiduous observation made with the utmost accuracy possible both of the *directions* and *durations* of meteors.

This attractive field of investigation has certainly been neglected in recent years, and the reason of this may perhaps

be found in the complications inseparable from it, in the need
of great patience and scrupulous care in observation, and the
necessity of gaining experience before the observer can feel a
reliance on his work, and draw safe conclusions. Meteors are
so fugitive, so diverse and erratic in their apparitions, as to be
quite beyond the scope of instrumental refinements. They must
necessarily be observed under many disadvantages. Positions
have to be fixed from very hurried and often imperfect impressions. But these drawbacks, formidable as they at first
appear, may be severally overcome by practice, by careful
regard for the conditions under which meteors are displayed,
and the marked differences of aspect induced by these conditions. When the observer has acquired a practical knowledge
he will proceed with confidence in his work, and avoid many
of the difficulties surrounding it.

In recording meteor-tracks for the purpose of discovering
the radiant-points, the chief feature in which precision is
essential is the *direction* of flight. A perfectly straight wand,
held in the hand for the purpose, should be projected upon the
path of every meteor directly it is seen, and then when the
eye has quickly noted the position and slope relatively to the
fixed stars near, it should be reproduced on the chart or celestial globe. The time, mag., estimated duration, and details of
appearance should be registered in a tabular form, with the
R.A. and Dec. of the beginning-point and probable radiant.
The end-point and length of path may be left until next day,
in order to save valuable time. The wand is a great assistance to the eye in retaining the approximate directions
and noting the places. If a meteor belongs to the slow,
trained class, or if it belongs to the swift, streak-leaving
order, the path may be very accurately noted, for the wand
can be adjusted to its direction before the meteor or its visible
offcome has died away. In the case of short, quick meteors,
devoid of either streaks or trains, and generally shooting from
radiants at high altitudes, they are more difficult to secure, as
they vanish before one may turn, and the observer must rely
upon the mere impression he received. But even these suc-
cumb to experience, and will be found to resolve themselves
into a number of sharply defined radiants scarcely less certain

than the positions derived from the streaked or trained meteors.

These positions are only to be fixed by the exercise of much cautious discrimination on the part of the observer, for the direction of the flight is not sufficient, alone, to indicate it. The visible aspect of the meteor has to be equally considered, for the place of its radiant imparts certain peculiarities to it which are rarely to be mistaken. First, *the astronomical position* of the radiant. If the radiant is at, or within 50° of, the Earth's apex (a point 90° preceding the Sun along the ecliptic, and towards which the Earth's motion is directed) the meteors generally leave streaks, especially the brighter ones, and move with great speed. They are usually white, exhibiting a high degree of incandescence. If the radiant is near the anti-apex or anywhere in the anti-apex half-sphere the meteors are streakless, they travel slowly or very slowly, and often leave trails of yellowish sparks. Bearing these facts in mind the region may be assigned in which any radiant is situated, if not the exact position of the radiant itself. If, say, on Aug. 10, at midnight a swift, streaked meteor is seen shooting from the Pleiades towards Aldebaran, just risen, the radiant is either in Musca, Triangulum, or Andromeda. But if the meteor is slow, with a train, then we must go further back in the direction of its flight, and seek the radiant in the S. or S.W. sky. If the motion is very slow, the radiant may be as far away as Aquila. Second, *the sensible position of the radiant.* A low radiant yields long-pathed meteors, characterized by slowness of speed and a flaky appearance either of the streaks or trains. A radiant near the zenith gives short, darting meteors, with rather dense streaks or trains. These nearly vertical meteors have a less extensive range of atmosphere to penetrate than the horizontal meteors, which are sometimes abnormally long. In the case of brilliant meteors, however, the paths occasionally extend over considerable arcs though the radiant may be high. Third, *the position of the radiant relatively to the path of a meteor.* If a meteor is close to its radiant its track is usually slow, and appears greatly foreshortened by the effects of perspective. It is travelling (approaching) nearly in the line of sight, and the streak or offcome of sparks is especially dense because it is seen through its entire depth; and the nucleus in such a case has a brushy diffused appearance. Such meteors often traverse sinuous, or curved paths of 2°, 3°, or 4°, and they are readily distinguishable from other meteors far from the radiants to which they belong.

A good method of tabulating meteor-tracks is that adopted by Lieut.-Col. Tupman in his catalogue published by the British Association in 1874. I have adopted the same form, and herewith append a copy of my register of a few isolated bright meteors observed in the autumn of 1890 :—

Date 1890.	G.M.T.	Mag.	Observed Path		Length of Path.	Dura-tion.	Appearance.	Pro-bable Radiant
			From R.A. Dec.	To R.A. Dec.				
	h m		° °	° °	°	sec.		° °
Oct. 17 ...	10 37	>1	219 +61	255 +65	16	3·5	V. slow, B. train.	204+56
19 ...	10 35	1	61½+26	44½+27½	15½	0·7	Swift, streak.	Orion.
19 ...	12 0	½ ☽	326 − 8	319−10	7	0·5	Swift, streak.	Orion.
25 ...	17 18	>♃	168 +34	180 +24	14½	0·8	Swift, streak.	Lynx.
26 ...	7 33	♃	329 +69	243 +51	42	4·0	Slow.	32+18
Nov. 1.....	7 1	>1	278 +49	244 +11½	46	6·0	Very slow.	50+15
1......	9 17	>1	345 +11	307 + 1½	39	4·0	Slow.	50+15
5......	10 40	>♀	28½−25	25½−29¼	5½	0·7	Swift, strk. 15 sec.	Taurus
16......	11 15	♃	274 +77	265½+67	10	1·5	Not very swift.	Auriga

The *duration of flight* is a most important element to esti-mate correctly, as it affords data wherewith the real velocity may be computed, and enables the nature of the orbit in which the meteor is moving to be definitely assigned. This feature is, however, one of the most difficult of all to derive with satisfactory precision. In the case of very slow meteors lasting several seconds, it is easy by means of a stop-watch, or by other methods, to get the times of flight within narrow limits of error, but the swifter class of meteors complete their visible trajectories in the fraction of a second, and are gone before any effort can be made to gauge their durations, so that a value has to be attributed which is little better than a mere guess.

Every adopted radiant-point should be based on at least five paths, unless the conditions are special, and these must show a very definite centre, and present family resemblances. It is often possible to detect a good centre from very few paths, when the radiant is low on the horizon, or when it occupies an isolated position.

In recording meteors the details of their appearances should also be appended to the paths. Foreshortened and crooked courses, fluctuations of brightness, halting motion, spark-trains, phosphorescent streaks, broken streaks, and other features must be invariably noted when observed, as likely to assist in fully comprehending these bodies. A streak will sometimes brighten up perceptibly after the head has died out.

One of the principal aims of future observers should be to ascertain the visible duration of meteor-showers, and the displacement or fixed position of the radiants during the period of their continuance. The Perseids seem to endure for forty-six nights (July 8–August 22) while the radiant moves from $3° + 49°$ to $76° + 57°$. The Lyrids also exhibit a shifting radiant, and it is highly probable some other showers are to be included in the same category. In investigating these, the observations of single nights should be kept separate, and the radiant determined from each set of paths. The positions when compared will then exhibit the rate and direction of the displacement. As to radiants which are apparently stationary * during long intervals, these should be closely observed. Are the centres of radiation, as successively determined, identical, allowing for the slight errors of observation? Are they continuously in operation, or intermittent? Meteors with motions in declination and near their radiants will be specially valuable in settling these questions, and if observed at more than one station will possess great significance. If it can be proved that a radiant is fixed and continuous during a few weeks, there can be no reason why it may not be stationary for a much more lengthy interval, unless the circumstances are exceptional.

* A list of these was published in the 'Monthly Notices,' vol. l. p. 466. See also 'Monthly Notices,' vol. xlv. pp. 93 *et seq.*

Though I have pointed out the urgency of noting the directions and durations of meteors, there are other features in such observations that must not be disregarded. If the paths are being recorded for the particular purpose of getting duplicate observations and calculating the heights, then it is desirable to note the beginning- and end-points of the flights as exactly as possible, for unless this is done the combined paths will show great discordances. Those who have acquired a familiar knowledge of the constellations will, however, experience little trouble in insuring accuracy in these records.

Observers, particularly those residing in towns, must be constantly on their guard against mistakes in identifying meteors from terrestrial objects such as fire-balloons and the various forms of pyrotechnic display. That such caution is necessary will be admitted when we read the two following letters, which were published in the 'Times' some years ago :—

" Sir,—

"A large meteor was seen to-night at 8.27, moving very slowly along the northern horizon, from west to east, at an altitude of about 8 deg. It was at least three times as brilliant as Venus, remaining visible for nearly five minutes, moving slower than any hitherto observed. I should be glad to receive observations made at more favourable stations. . .

" I remain, Sir, your obedient Servant,

" Thomas Crumplen.

" Mr. Slater's Observatory,
Euston Road,
August 10th."

" Sir,—

" The 'large meteor' seen by Mr. Crumplen on Monday evening at 8.27, three times as brilliant as Venus, and moving from west to east, was a fire-balloon sent up shortly after 8 o'clock from the Eton and Middlesex Cricket Ground, Primrose Hill, as a *finale* to some athletic sports which had taken place during the afternoon.

" I am, Sir, your obedient Servant,

" B. C. C.

" St. John's Wood,
August 12th."

In concluding this chapter I may briefly mention that an old idea concerning meteors was that they originated gales of wind, and that, in fact, they were the usual precursors of stormy weather. This belief is thus expressed in Dryden's ' Virgil":—

> " Oft shalt thou see, ere brooding storms arise,
> Star after star glide headlong down the skies,
> And, where they shot, long trails of lingering light,
> Sweep far behind, and gild the shades of night."

CHAPTER XVI.

THE STARS.

> " Ten thousand suns appear
> Of elder beam ; which ask no leave to shine
> Of our terrestrial star, nor borrow light
> From the proud regent of our scanty day."
>
> BARBAULD.

THE planetary observer has to accept such opportunities as
are given him ; he must use his telescope at the particular
seasons when his objects are well presented. These are
limited in number, and months may pass without one of them
coming under favourable review. In stellar work no such
irregularities can affect the progress of observations. The
student of sidereal astronomy has a vast field to explore, and
a diversity of objects of infinite extent. They are so various
in their lustre, in their grouping, and in their colours, that
the observer's interest is actively retained in his work, and we
often find him pursuing it with unflagging diligence through
many years. No doubt there would be many others employing
their energies in this rich field of labour but for the unin-
teresting character of star-disks, which are mere points of
light, and therefore incapable of displaying any detail. Those
who study the Sun, Moon, or planets have a large amount
of surface-configuration to examine and delineate, and this is
ever undergoing real or apparent changes. But this is wholly
wanting in the telescopic images of stars, which exhibit a
sameness and lack of detail that is not satisfying to the tastes
of every observer. True there are some beautiful contrasts

of colour and many striking differences of magnitude in double stars; there are also the varying position and distance of binary systems, the curious and mysterious fluctuations in variable stars, and some other peculiarities of stellar phenomena which must, and ever will, attract all the attention that such important and pleasing features deserve. And these, it must be conceded, form adequate compensation for any other shortcomings. The observer who is led to study the stars by comparisons of colour and magnitude or measures of position, will not only find ample materials for a life-long research, but will meet with many objects affording him special entertainment. And his work, if rightly directed and accurately performed, will certainly add something to our knowledge of a branch in which he will certainly find much delectation.

Greek Alphabet.—The amateur must, at the outset of his career, thoroughly master the Greek alphabet. This will prevent many time-wasting references afterwards, and avoid the doubt and confusion that must otherwise result. The naked-eye stars in each constellation have Greek letters affixed to them on our celestial globes and star-maps.

α	Alpha	ν	Nu
β	Beta	ξ	Xi
γ	Gamma	o	Omĭcron
δ	Delta	π	Pi
ϵ	Epsĭlon	ρ	Rho
ζ	Zēta	σ	Sigma
η	Eta	τ	Tau
θ	Theta	υ	Upsĭlon
ι	Iota	ϕ	Phi
κ	Kappa	χ	Chi
λ	Lambda	ψ	Psi
μ	Mu	ω	Omĕga.

The letters are applied progressively to the stars (generally according to brightness) in each constellation. The 1st-mag. stars frequently have a duplicate name. Thus *a* Leonis is also known as Regulus, and *a* Canis Majoris as Sirius, the Dog-star.

Learning the Names of the Stars.—A knowledge of the

stars as they are presented in the nocturnal sky may be regarded as the entrance to the more advanced and difficult branches of the science, and forms the young observer's introductory lesson. When he has learnt a few of the principal constellations, and can point them out to his friends, he already begins to feel more at home with the subject, and regards it with a different eye to what he did before when the names and configurations of the stars were alike unknown to him. He no longer views the heavens as a mysterious assemblage of confusing objects, for here and there he espies certain well-known groups always preserving the same relative positions to each other. The unconscious gaze he formerly directed to the sky has given way to the intelligent look of recognition with which he now surveys the firmament.

An acquaintance with the leading constellations, and with the names or the letters of the brighter stars in each, becomes very important in some departments of observation, and various methods have been suggested as likely to impress the positions and names on the memory. The beginner must first be content to get familiar with a few of the brighter stars, and make these the base for extending his knowledge. The objects are so numerous that it is impossible his primary attempts can be anything like complete. He must advance step by step in his survey, and feel his way cautiously, setting out from certain conspicuous stars with which he has already become conversant. A lantern and a series of star-maps are the only aids required, and with these he ought to make satisfactory progress. The stars as they are seen in the sky may be compared with those figured in the maps, and their names and the constellations in which they lie may then be identified. It is an excellent plan as conducing to fix the positions indelibly in the memory to construct maps from personal observation, and to compare these afterwards with the published maps for identification of the constituent stars. This plan, if repeated several times, has the effect of impressing the positions of the leading stars forcibly upon the observer's mind.

It is not intended to give, in this place, any details as to the places or distribution of the stars. Without diagrams, such a description could not be made readily intelligible. To those, however, who are commencing their studies, I would recommend the northern sky as the most suitable region to aid their initiatory efforts. For

> " He who would scan the figured sky
> Its brightest gems to tell,
> Must first direct his mind's eye north
> And learn the Bear's stars well."

The seven bright stars of Ursa Major are familiar to nearly everyone. Two of them, called the Pointers, serve to direct the eye to the Polar star, which, though not a brilliant one, stands out prominently in a region comparatively bare of large stars. It is important to know the Polar star, as it is situated near the centre of the apparent motion of the firmament. When the student has assured himself as to the northern stars

Fig. 60.

The constellation Orion.

he will turn his attention southwards, and recognize the beautiful Orion and the curious groups in Taurus. He will also observe, much further east, the well-known sickle of Leo, and in time become acquainted with the many other constellations that make the winter sky so attractive.

U

The Constellation Figures.—The observer will soon realize
that the creatures after which the constellations have been
named bear no resemblance to the configuration of the stars
they represent. If we look for a Bear amongst the stars of
Ursa, for a Bull amid the stars of Taurus, or for a flying Swan
in the stars of Cygnus we shall utterly fail to find it. The
names appear to have been originally given, not because of
individual likenesses between them and the star-groups to
which they are applied, but simply on account of the necessity of dividing the sky into parts, and giving each a distinguishing appellation, so that it might be conveniently referred
to. There were pressing needs for a system of stellar nomenclature, and the plan of grouping the stars into imaginary
figures was the one adopted to avoid the confusion of looking
upon the sky as a whole. There are some who object to the
method of the Chaldean shepherds because the series of
grotesque figures on our star-maps and globes bear no natural
analogies. But it would be unwise to attempt an innovation
in what has been handed down to us from the myths of a
remote antiquity, for

> " Time doth consecrate,
> And what is grey with age, becomes religion."

Means of Measurement.—A micrometer becomes an indispensable instrument to those who make sidereal observations
of an exact character. Without such means it will be impossible to determine either positions or distances except by mere
estimation, and this is not sufficiently precise for double-star
work. With a reliable micrometer * excellent results may
be obtained, especially with regard to the varying angles of
binary systems. Frequent remeasurement of these is desirable for comparison with the predicted places in cases
where the orbits have been computed. In this department
of astronomy the names of Herschel, South, Struve, Dawes,
Dembowski, Burnham, and others are honourably associated,
and it is notable that refracting-telescopes have accomplished

---

* There are several forms of this instrument: for particulars of construction and use the reader is referred to Thornthwaite's ' Hints on
Telescopes,' and Chambers's ' Astronomy,' 4th ed. vol. ii.

nearly the whole of the work. But reflectors are little less
capable, though their powers seem to have been but rarely
employed in this field. Mr. Tarrant has lately secured a
large number of accurate measures with a 10-inch reflector
by Calver, and if care is taken to secure correct adjustment
of the mirrors, there is no reason why this form of instrument

Fig. 61.

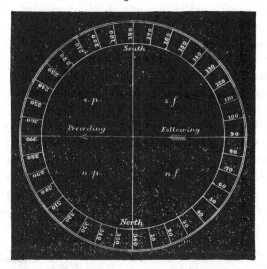

Diagram illustrating the Measurement of Angles of Position.

(In measuring angles of position the larger star is always understood as
central in the field. The north point is zero, and the angles are
reckoned from this point towards the east. If a star has a faint
component lying exactly east or following it, then the angle is 90°;
if the smaller star is south, the angle is 180°; and so on.)

should not be nearly as effective as its rival. Mr. Tarrant
advises those who use reflectors in observing double stars " to
test the centering of the flat at intervals during the observations, as the slightest shift of the large mirror in its cell
will frequently occasion a spurious image which, if it by
chance happens to fall where the companion is expected to
be seen, will often lead to the conclusion that it has been
observed. In addition to this, any wings or the slightest

flare around a bright star will generally completely obliterate every trace of the companion, especially if close and of small magnitude, and such defects will in nine cases out of ten be found to be due to defective adjustment. Undoubtedly for very close unequal pairs the refractor possesses great advantages over a reflector of equal aperture ; in the case of close double stars the components of which are nearly equal there appears to be little, if any, difference between the two classes of instruments ; while for any detail connected with the colour of stars the reflector certainly comes to the fore from its being perfectly achromatic." These remarks from a practical man will go far to negative the disparaging statements sometimes made with regard to reflectors and stellar work, and ought to encourage other amateurs possessing these instruments to take up this branch in a systematic way.

Dividing Power.—This mainly depends upon the aperture, and it was made the subject of careful investigation and experiment by Dawes, who found that the diameters of the star-disks varied inversely as the aperture of the telescope. Adopting an inch as the standard, he ascertained that this aperture divided stars of the sixth magnitude 4″·56 apart, and on this base he constructed the following table :—

Aperture in inches.	Least separable distance.	Aperture in inches.	Least separable distance.
1·0	4″·56	6·5	0″·70
1·6	2·85	7·0	0·65
2·0	2·28	7·5	0·61
2·25	2·03	8·0	0·57
2·5	1·82	8·5	0·536
2·75	1·66	9·0	0·507
3·0	1·52	9·5	0·480
3·5	1·30	10·0	0·456
3·8	1·20	12·0	0·380
4·0	1·14	15·0	0·304
4·5	1·01	20·0	0·228
5·0	0·91	25·0	0·182
5·5	0·83	30·0	0·152
6·0	0·76		

Dallmeyer, the optician, confirmed these values by remarking :—" In all the calculations I have made, I find that 4·33 divided by the aperture gives the separating power. Thus, 4·33 inches separates 1"." But a good deal depends upon the character of the seeing and upon other conditions. A large aperture will sometimes fail to reveal a difficult and close *comes* to a bright star when a smaller aperture will succeed. This is due to the position of the bright diffraction-ring, which in a large instrument may overlap the faint companion and obscure it, while in a small one the ring falls outside and the small star is visible*. Dawes concluded that " tests of separation of double stars are not tests of excellence of figure," and he gave much valuable information with regard to micrometers and double-star observations generally .in the 'Monthly Notices,' vol. xxvii. pp. 217–238. This paper will well repay attentive reading.

Number of Stars.—In the northern hemisphere there are about 5000 † stars perceptible to the naked eye. This is less than an observer would suppose from a casual glance at the firmament, but hasty ideas are often inaccurate. The scintillation of the stars and the fact that many small stars are momentarily glimpsed but cannot be held steadily have a tendency to occasion an exaggerated estimate of their numbers. Authorities differ as to the total of naked-eye stars. Sir R. S. Ball says " the number of stars which can be seen with the unaided eye in England may be estimated at about 3000." Gore gives 4000. Backhouse mentions 5600 as visible in the northern hemisphere. Argelander, who has charted 324,188

---

* Mr. George Knott, of Cuckfield, mentions that the radius of the first bright diffraction-ring of a stellar image, for a 7⅓-inch aperture, is 1"·01, and for one of 2 inches 3"·70 ('Observatory,' vol. vi. p. 19 ; see also vol. i. pp. 107 and 145). Mr. Dawes is quoted as giving 1"·25 for a 7-inch, 1"·61 for a 5½-inch, and 3"·57 for a 2·4-inch. These figures exceed the theoretical values, if the latter are adopted from Sir G. B. Airy's ' Undulatory Theory of Optics,' where, for mean rays, we have :—

Radius of object-glass in inches × radius of bright ring in seconds = 3·70.

† The number visible to different persons varies according to eyesight. Some observers see thirteen or fourteen stars in the Pleiades, while others cannot discern more than six or seven.

stars between 2° S. of the equator and the N. pole, gives the following numbers of stars up to the 9th magnitude :—

1st.	2nd.	3rd.	4th.	5th.
20	65	190	425	1100

6th.	7th.	8th.	9th.
3200	13,000	40,000	142,000

With every decrease in magnitude there is a great increase in numbers, and if this is extended to still smaller magnitudes down to the 15th or 16th we can readily understand that there exist vast multitudes of fainter stars. Paul Henry, of the Paris Observatory, says there are about 1,500,000 stars of the 11th mag., and Dr. Schönfield, of Bonn, gives 3,250,000 as of the $11\frac{1}{2}$ mag. It is probable that by means of photography and the largest telescopes considerably more than 50 millions. of stars may be charted, but this is a mere approximation. Roberts has photographed 16,206 stars within an area of four square degrees in a very rich region of the Galaxy near η Cygni, and Gore computes that were the distribution equal to this over the whole firmament the number of stars would reach 167 millions. He also remarks that in the Paris photographs of the Pleiades, 2326 stars are shown in a space covering about three square degrees, and this gives for the entire sky a total of 33 millions. It is, however, manifest that unusually plentiful spots in the heavens cannot be accepted as affording a criterion of the whole.

Magnitudes.—According to Argelander's figures, above quoted, each magnitude exhibits a rise of about 300 per cent. But authorities rarely agree as to scale, as the following comparison between Sir J. Herschel and Struve will show:—

H.	S.	H.	S.
4·0	3·6	8·0	7·3
4·5	4·1	8·5	7·7
5·0	4·6	9·0	8·1
5·5	5·05	9·5	8·5
6·0	5·5	10·0	8·8
6·5	5·95	10·5	9·1
7·0	6·4	11·0	9·3
7·5	6·85	11·5	9·6

H.	S.	H.	S.
12·0	9·8	15·0	10·87
12·5	10·0	16·0	11·13
13·0	10·18	17·0	11·38
13·5	10·36	18·0	11·61
14·0	10·54	19·0	11·82
14·5	10·71	20·0	12·00

Argelander's magnitudes come between those of Herschel and Struve. Such disagreements are perplexing to observers, and it is fortunate that in regard to the naked-eye stars we are now furnished with a more consistent and accurate series of magnitudes. Photometric determinations of the light of 4260 stars not fainter than the 6th mag., and between the N. pole and 30° S. declination, were made at Harvard College Observatory, and similar measures of 2784 stars between the N. pole and 10° S. declination were effected at the Oxford University Observatory, and the results published in 1885. The two catalogues are in very satisfactory agreement, the accordances within one tenth of a mag. being 31 per cent., within one quarter of a mag. 71 per cent., and within one third of a magnitude 95 per cent. The photometers used in the two independent researches were constructed on very different principles, and the substantial agreement in the results indicates that " a great step has been accomplished towards an accurate knowledge of the relative lustre of the stars " ('Monthly Notices,' vol. xlvi. p. 277).

The Milky Way.—On dark nights when the Moon is absent and the air clear, a broad zone of glimmering, filmy material is seen to stretch irregularly across the heavens. It may be likened to a milky river running very unevenly amongst the constellations, and showing many curves and branches along its course. On very favourable occasions the unaided eye glimpses many hundreds of glittering points on this light background. A field-glass reveals some thousands, and shows that it is entirely composed of stars the blended and confused lustre of which occasions that track of whiteness which is so evident to the eye. In a good telescope stars and star-dust exist in countless profusion, and great diversity is apparent in their numbers and manner of grouping. In certain regions the stars are concentrated into swarms, and the sky is aglow with them ; while in others there are very few, and dark cavernous openings offer a striking contrast to the silvery sheen of surrounding stars. There are many of these void spaces in Scorpio, and a circular one in Sagittarius R.A. 17^h 56^m, Dec.—27° 51' has been particularly remarked. These inequalities of grouping may be easily recognized with the naked eye, especially in Cygnus, where bright star-lit regions frequently alternate with dark void spaces. In the southern sky there is a noteworthy instance. Near the brilliant stars of Crux and Centaurus and closely surrounded by the Milky Way there is a large black vacancy very obvious at a glance, and so striking to ordinary observers that it is known as the " Coalsack," a name applied to it by the early navigators of the southern seas.

The course of the Milky Way may be described generally as flowing through Auriga, the club of Orion, feet of Gemini, western part of Monoceros, Argo Navis, Crux, feet of Centaurus, Circinus, Ara, where it separates into two branches, the western of which traverses the northern part of the tail of Scorpio, eastern side of Serpens, Taurus Poniatowski, Anser, and Cygnus. The eastern branch crosses the tail of Scorpio, the bow of Sagittarius, Antinous, Aquila, Vulpecula, and then enters Cygnus, where it reunites with the other branch. It thence passes through Cepheus, Cassiopeia, Perseus, and enters Auriga. In breadth it varies greatly, being in some places only 4° or 5°, whereas in others it reaches 20°. It is, of course, best visible when twilight is absent, but it is sometimes very plain, even at midsummer, for at this season some of its more conspicuous sections are favourably placed for observation. It is supposed that fully nine tenths of the total number of stars in the firmament are included within the borders of the Milky Way.

Some of the ancient philosophers, including Democritus, formed just conceptions as to the real nature of this appearance. Though they lacked instruments wherewith to observe the stars forming it, they yet saw them with the eye of reason. But very vague and incorrect notions prevailed in early times, when superstition was rife, as to many celestial phenomena. Some of the ancient poets and learned men refer to the Galaxy as the path by which heroes ascended to heaven. Thus we read in Ovid :—

> " A way there is in heaven's extended plain,
> Which when the skies are clear is seen below,
> And mortals, by the name of Milky, know ;
> The ground-work is of stars, through which the road
> Lies open to great Jupiter's abode."

Scintillation of the Stars.—The rapid variations of light known as the "twinkling" of the stars received notice from many ancient observers, including Aristotle, Ptolemy, and others, and they severally endeavoured to account for it, but not in a manner altogether satisfactory. At low altitudes bright stars exhibit this twinkling or scintillation in a striking degree, but it is much less perceptible in stars placed at considerable elevations. Sirius, the brightest star in the sky, is a noted twinkler. His excessive lustre and invariably low position are conditions eminently favourable to induce this effect. But the planets seldom exhibit scintillation in a very marked degree. The light of Jupiter and Saturn is steady, even when these planets are close to the horizon. Mercury, however, twinkles most obviously, and Venus and Mars, when low down, are often similarly affected, especially in stormy weather when the air is much disturbed. Hooke, in 1667, concluded that the scintillation was due " to irregular refractions of the light of the stars by differently heated layers of atmosphere." M. Arago said it arose " from the peculiar properties possessed by the constituent rays of light, of moving with different velocities through the strata of the atmosphere, and of producing what are called interferences." More recently, M. Montigney has conducted some interesting researches into this subject, and he believes " that not only is twinkling caused, to a great extent, by the deviations of portions of a star's light altogether away from us by variable layers of atmosphere, but it is also affected, both in frequency and in the colours displayed, by the nature of the light emitted by the individual star." The planets are little subject to scintillation, as they present disks of sensible size, and thus are enabled to neutralize the effect of atmospheric interferences.

It is curious, however, that the steadiness of telescopic images does not appear to be much improved at high altitudes, and that the phenomenon of scintillation still operates powerfully as observed from mountainous stations. In February 1888, Dr. Pernter, of the Vienna Academy of Sciences, found "that the scintillation of Sirius was actually greater at the top of Sonnblick, 10,000 feet high, than it was at the base of the mountain, and he formed the opinion that scintillation has its origin in the *upper* strata of the atmosphere and not in the lower as usually assumed." It would appear from this that lofty situations do not possess all the advantages claimed for them in regard to the employment of large telescopes.

Star-Disks.—The stars as observed in telescopes are shorn of the false rays apparent to the naked eye, and they are seen with small spurious disks. That the disks are spurious is evident from the fact that the larger the telescope employed, the smaller the star-disks become. And moreover, when a star is occulted by the Moon, it disappears instantaneously. There is no gradual diminution of lustre ; the star vanishes with great suddenness. Bright stars, like Aldebaran or Regulus, have been watched up to the Moon's limb, and observers have been sometimes startled at the abruptness with which they were blotted out. An appreciable disk could not be withdrawn in this instantaneous manner ; it would exhibit a perceptible decadence as the Moon increasingly overlapped it, but no such appearance is observed. On the occasion of the occultation of Jupiter on Aug. 7, 1889, the planet's diameter was 41″·4, and the disappearance occupied 85 seconds. Now had Aldebaran or Regulus a real disk of only 1″ it would prevent their sudden extinctions, and their disappearances would be spread over perceptible though short intervals of time[*]. But there is every reason to conclude that the actual disks are to be represented by a small fraction of 1″, so that the largest instrument and the highest powers fail to reveal it. In this connection, Mr. Gore, in his ' Scenery of the

[*] About 2 seconds. Sir W. Herschel found the diameter of *a* Lyræ with a power of 6450 to be 0″·3553. Tycho Brahe, before the invention of telescopes, estimated the diameter of Sirius as 120″. J. D. Cassini, with a telescope 35 feet long, found the diameter of the same star 5″.

Heavens,' p. 152, says :—" Let us take the case of *a* Centauri, which is, as far as is known at present, the nearest fixed star to the Earth. The distance of this star is about 25 billions of miles. From comparisons made between its light and the Moon, it has been found that its intrinsic brilliancy must be about four times that of the Sun. Supposing its greater lustre is due to its greater size—a not improbable supposition—it would subtend, if placed at the Sun's distance, an angle twice as great, or about 1°, and hence we find that the angle subtended at its distance of 25 billions of miles would be about $\frac{1}{76}$th of a second of arc, which the most powerful telescope yet constructed would be incapable of showing as a visible disk."

Distance of the Stars.—The distances of the outer planets Uranus and Neptune, mentioned in an earlier chapter of this work, are sufficiently large to amaze us ; but the distances of the stars may be said to be relatively infinite. For many years the problem of stellar distances defied all attempts to resolve it. At length, in 1838–39, Bessell, Henderson, and Struve obtained results for three stars—viz. 61 Cygni, *a* Centauri, and *a* Lyræ,—which practically settled the question. More recent measures of stellar parallax, while correcting the earlier values, have virtually corroborated them ; though the figures adopted can only be regarded as approximations, owing to the difficult and delicate nature of the work. The binary star *a* Centauri appears to be the nearest of all ; it has a parallax of $0''\cdot75$, and its distance from us is equal to 275,000 times the distance of the Sun. Light traversing space at the rate of 187,000 miles per second would occupy $4\frac{1}{3}$ years in crossing this interval. In the Northern hemisphere 61 Cygni is the nearest star, with a parallax of $0''\cdot44$ and a distance of about 470,000 times the Sun's distance. Light would take more than seven years in reaching us from this star. *a* Lyræ has a parallax of $0''\cdot15$, equal to nearly 22 light-years. *a* Crucis shows a very small parallax ($0''\cdot03$), and its distance is excessively remote—equal to about 108 light-years !

Proper Motion of Stars.—A very slight motion affects the places of many of the so-called fixed stars. This must, after the lapse of long intervals of time, materially alter the con-

figuration of the constellations. But the change is a very
gradual one, and must operate through many centuries before
its effects will become appreciable in a general way. The
greatest proper motion yet observed is that in regard to two
small stars (one in Ursa Major and the other in Piscis
Australis), which amounts to about 7″ annually. Another
motion has been recognized, viz. in the line of sight. Dr.
Huggins made the initiatory efforts in this research by
measuring the displacement of the F line in the spectrum of
Sirius. The work has been actively pursued at the observatories of Greenwich and Rugby, and with interesting
results. While certain stars exhibit a motion of approach,
others display a motion of recession. Thus Vega, Arcturus,
and Pollux are approaching us at the rate of about 40 miles
per second ; while Rigel is receding at the rate of 17 miles
per second, Castor at the rate of 19, Regulus 14, Betelgeuse 25,
and Aldebaran 31. Sirius, in the years from 1875 to 1878,
was receding from us at the rate of 22 miles per second ; but
this decreased in subsequent years, and in 1884–85 the star
was approaching with a motion of about 22 miles per second.
In 1886 and 1887 this rate was increased to about 30 miles
per second, as observed both at Greenwich and Rugby.
This confirms the idea that Sirius is moving in an elliptical
orbit. Similar observations, in regard to the variable star
Algol, have revealed that changes of velocity are connected
with its changes of lustre. Before minimum the star recedes
at the rate of $24\frac{1}{2}$ miles per second, while after minimum the
star approaches with a speed of $28\frac{1}{2}$ miles per second (' Monthly
Notices,' vol. l. p. 241).

Double Stars and Binary Systems.—Telescopic power will
often reveal two stars where but one is seen by the naked eye.
Sometimes the juxtaposition of such stars is merely accidental ; though they are placed nearly in the same line of
sight the conjunction is an optical one only, and no connection apparently subsists between them. In other cases,
however, pairs are found which have a physical relation, for
one is revolving round the other ; and these are termed *binary*
stars. Sir W. Herschel was the first to announce them, from
definite observations, in 1802. Of double stars more than

10,000 are now known ; many of these are telescopic, but the list includes some fine examples of naked-eye stars.

Double stars are excellent telescopic tests. A very close pair affords a good criterion as to the defining capacity of an instrument ; while a pair more widely separated and of greatly unequal magnitude, like that of *α* Lyræ, offers a test of the light-grasping power. But in these delicate observations, as, indeed, in all others, the character of the seeing exercises an important and variable influence. A double star that is well shown on one night becomes utterly obliterated on another, owing to the unsteadiness and flaring of the image. On such occasions as the latter one is reminded of the " twitching, twirling, wrinkling, and horrible moulding " of which Sir John Herschel complained, and which unfortunately forms a too common experience of the astronomical observer. A close double, of nearly equal magnitudes, requires a steady night,

Fig. 62.

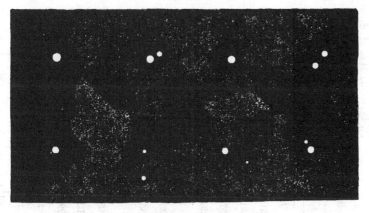

Double Stars.

β Orionis.	γ Leonis.	α Ursæ Minoris.	γ Virginis.
δ Cygni.	γ Arietis.	γ Andromedæ.	δ Serpentis.

such as is suitable for planetary details ; but a wide double consisting of a bright and a minute star rather needs a very clear sky than the perfection of definition. Certain doubles, such as *θ* Aurigæ, δ Cygni, and *ζ* Herculis, are often more easily seen in twilight than on a dark sky ; and some expe-

List of Double Stars.

[Abbreviations in col. 9:—β., Burnham; T, Tarrant; S., Schiaparelli; L., Leavenworth; E., Engelmann; P., Perrotin; HΣ., H. Struve; M., Maw.]

No.	Name of Star.	Position, 1890. R.A.	Position, 1890. Dec.	Mags.	Position-Angle.	Distance.	Epoch.	Observer.	Notes.
		h m	° ′		°	″			
1.	δ Equulei	21 9·1	+ 9 34	4½ 5	189·9	0·25	1857·7	β.	Most rapid binary known. Period 11¾ years (Wrublewsky). Disc. 1852 by O. Struve.
2.	Piazzi 109	1 51·0	+ 1 20	7 7	206·3	0·28	1888·1	S.	An excessively close and difficult object. Binary.
3.	β Delphini	20 32·4	+14 13	3½ 5½	310·1	0·29	1888·6	β.	A rapid binary. Period 26 years (Doubjago). Disc. 1873 by Burnham.
4.	γ² Andromedæ	1 57·1	+41 48	5 6	277·6	0·35	1884·8	L.	Distance in Oct. 1889 less than 0″·1, and very difficult with 36-inch (Burnham).
5.	γ Coronæ Bor.	15 38·1	+26 39	4 7	126·6	0·38	1887·5	S.	A close binary. Period 95½ years (Doberck).
6.	55 Tauri	4 13·6	+16 16	6½ 8	76·4	0·43	1887·6	S.	Colours greenish-white and purple. A binary. Difficult object with a 10-inch.
7.	λ Cassiopeiæ	0 25·7	+53 55	6½ 6½	146·9	0·45	1887·3	T.	Another close binary. Distance of components shows little change.
8.	ζ Boötis	14 35·9	+14 12	4 4	293·4	0·51	1887·5	S.	A binary pair, of equal mags. Period 127 years (Doberck).
9.	42 Comæ Bor.	13 4·7	+18 7	5½ 6	189·6	0·55	1889·1	L.	A close binary, of short period; about 25¾ years. Disc. in 1827 by O. Struve.
10.	λ Cygni	20 43·1	+36 8	5 7½	70·6	0·63	1888·8	HΣ.	A binary. The distance between the components is increasing.
11.	η Coronæ Bor.	15 18·7	+30 41	5½ 6	178·5	0·63	1886·5	T.	A well-known binary, of short period; 41½ years (Doberck).
12.	ω Leonis	9 22·6	+ 9 32	5½ 7	96·8	0·70	1889·1	L.	A close pair, but not difficult. Binary. Period 114½ years (Doberck).
13.	15 Lyncis	6 47·8	+58 34	5 6	5·9	0·77	1890·3	M.	A probable binary, the position and distance exhibiting a gradual increase.
14.	ι Orionis	5 1·9	+ 8 21	5½ 7	193·9	0·99	1889·0	L.	Triple. A low power shows many stars here.
15.	ζ Cancri, A.B.	8 5·9	+18 0	5 6	40·3	1·05	1889·2	L.	A triple star. A.C. Pos. 134°·4; Dist. 5″·36; Mag. 7; 1878·3 (Hall).

No.	Star	R.A.	Dec.	Mag.	Mag.	P.A.	Dist.	Epoch	Authority	Remarks
16.	ν Scorpii, A.B.	16 5·6	−19 10	4	7	9·3	1·08	1886·5	T.	A quadruple star, forming one of the finest systems in the sky.
17.	π Cephei	23 4·4	+74 47	5	7½	32·5	1·16	1888·7	HΣ.	Binary. Becoming more difficult with decrease of distance. Yellow and purple.
18.	ε Arietis	2 52·9	+20 54	5½	6	202·2	1·28	1889·7	L.	Distance increasing. Good dividing-test for a 4-inch aperture (T.).
19.	λ Ophiuchi	16 25·4	+ 2 13	4½	5⅘	42·6	1·55	1888·4	L.	Binary, but period not yet ascertained with accuracy. Yellow and bluish.
20.	ζ Herculis	16 37·1	+31 48	3	6½	65·8	1·68	1890·7	M.	A fine, rather close binary. Period 34½ years (Doberck). Single in 1865. Yellow and red.
21.	ξ Ursæ Maj.	11 12·3	+32 9	4	5	222·7	1·63	1889·3	S.	One of the first-computed binaries. Period 63 years (Breen). Excellent object.
22.	δ Cygni	19 41·5	+44 52	3	8	317·7	1·66	1885·5	T.	A well-known binary. Period 376·7 years (Gore). Test for 4½-inch. Pale yellow and sea-green.
23.	33 Orionis	5 25·5	+ 3 12	5	6	32·8	1·81	1887·1	T.	Just visible in a 3-inch. White and pale blue.
24.	θ Aurigæ, A.B.	5 52·2	+37 12	3	8	2·5	1·98	1885·1	T.	A similar pair to δ Cygni, though the distance is wider.
25.	70 Ophiuchi	18 0·0	+ 2 32	4	6	348·7	2·16	1889·3	β.	Binary. Period nearly 88 years (Gore). Good object for a 3-inch. Yellow and purple.
26.	ι Leonis	11 18·2	+11 8	4½	7½	62·0	2·56	1889·2	L.	Binary; but distance shows little variation since 1839. Yellowish and blue.
27.	ε Boötis	14 40·2	+27 32	3	5½	328·1	2·88	1885·4	T.	A very interesting object, and visible in a small instrument.
28.	α Scorpii	16 22·7	−26 11	1	8	271·7	2·92	1880·0	β.	This pair forms an atmospheric rather than an optical test.
29.	γ Ceti	2 37·6	+ 2 46	3	7	280·7	2·94	1883·9	P.	A binary system. Test for a 2½-inch. Yellow and blue.
30.	α Piscium	1 56·3	+ 2 14	5	6	321·9	3·03	1886·9	T.	A probable binary, but since 1831 not much change in position or distance.
31.	ζ Aquarii	22 23·1	− 0 35	4	4	325·8	3·08	1889·9	L.	A fine binary, with very long period. 1625 years (Doberck).
32.	ε¹ Lyræ	18 40·7	+39 34	4½	6½	15·3	3·24	1877·4	Doberck.	These stars form a wide double (distance 3′ 27″), just separable by the naked eye. A 2½-inch shows a fine double-double.
33.	ε² Lyræ	18 40·7	+39 30	5	5	137·6	2·45	1877·4	Hall.	A 4-inch reveals three faint stars between.

List of Double Stars (continued).

No.	Name of Star.	Position, 1800. R.A.	Position, 1800. Dec.	Mags.		Position-Angle.	Distance.	Epoch.	Observer.	Notes.
		h m	° ′			°	″			
34.	ε Hydræ	8 41·0	+ 6 49	4	7	226·5	3·16	1889·1	β.	A new *comes*, Pos. 154°·4; Dist. 0″·26; Mag. 6; 1889; 36-inch, power 3300! β.
35.	γ Leonis, A.B.	10 13·9	+20 24	2	4	114·6	3·51	1889·3	β.	A fine binary. Period 407 years (Doberck). Readily seen in a 3-inch.
36.	δ Serpentis	15 29·6	+10 55	3	5	189·9	3·52	1886·6	Ball.	Probably binary. Fine object in small instruments.
37.	α Canis Maj.	6 40·3	−16 34	1	10	359·7	4·19	1890·3	β.	Brilliant binary. Period 58·5 years (Gore). Colours white and yellow.
38.	α Herculis	17 9·6	+14 31	3	4½	114·5	4·58	1885·5	T.	A splendid object. Orange and bluish green.
39.	η Cassiopeiæ	0 42·4	+57 14	4	8	184·7	4·76	1888·3	M.	Binary. Period 195 years (Gruber). Difficult object for 2¼-inch (Johnson).
40.	γ Virginis	12 36·1	− 0 51	3	3	153·9	5·45	1889·3	L.	Well-known binary. Period 182 years (J. Herschel). Single in 1836.
41.	α Geminorum	7 27·6	+32 8	2	3	229·4	5·68	1889·2	L	Very fine object. Binary; Period doubtful (Mädler 232 years, Doberck 1001 years).
42.	π Boötis	14 35·6	+16 54	4	6	104·3	6·04	1885·4	T.	This pair has exhibited little change in pos. or dist. since 1781.
43.	α² Capricorni, A.B.	20 11·9	−12 53	3	15	149·7	6·30	1879·7	β.	Good light-test for 6 inches. Companion double; pos. 240°, dist. 1″·5.
44.	δ Geminorum	7 13·5	+22 11	3½	9	207·2	6·98	1886·1	T.	Rather wide pair of unequal mags. Difficult with small apertures.
45.	γ Arietis	1 47·5	+18 45	4½	5	178·3	8·78	1886·9	T.	A fine, easy object. Discovered in 1664 by Hooke.
46.	ι Ursæ Maj.	8 51·7	+48 28	3	12	356·7	9·56	1883 4	E.	Well seen in a 4-inch, powers 80 and 130. Good light-test.
47.	β Orionis	5 9·3	− 8 20	1	9	202·0	9·61	1887·2	T.	A fine object for small instruments. Visible in a 2-inch refractor.
48.	γ′ Andromedæ	1 57·1	+41 48	3	6	62·6	10·50	1876 0	Hall.	A splendid pair, stationary in relative positions (see no. 4).

No.	Name	R.A.	Dec.	Mag. 1	Mag. 2	Pos.	Dist.	Epoch	Obs.	Remarks
49.	γ Delphini	20 41·6	+15 44	4	6	271·2	11·35	1879·7	Hall.	Estimates of the colour of this pair differ, and change is inferred.
50.	σ Orionis, A.D.	5 33·2	− 2 40	4	10½	236·8	11·62	1875·2	......	Multiple. Fine group here. Schröter saw 12 stars, Struve 18.
51.	β Scorpii	15 59·0	−19 30	2	5½	26·7	12·72	1879·7	β.	The brighter star is a close double; Pos. 87°, Dist. 0″·73 (Burnham).
52.	ζ Ursæ Maj.	13 19·5	+55 30	2	4	150·5	14·38	1886·2	T.	Fine object for small instruments. Other stars in the field.
53.	α Centauri	14 32·1	−60 23	1	2	202·9	17·12	1888·6	S.	A fine southern binary with Period of 80·3 years (Elkin).
54.	α Ursæ Min.	1 18·5	+88 43	2	9	210·1	18·60	......	......	Good test for a 2-inch. Dawes saw it with 1 3/10-inch, Ward with 1¼ inch.
55.	61 Cygni	21 2·0	+38 12	5	6	121·0	20·58	1887·7	S.	Probably a binary of long period (782½ years, Peters; 1159 years, Mann).
56.	33 Arietis	2 34·3	+26 35	5	8	0·3	29·76	1879·7	β.	A distant and easy pair in small instruments.
57.	β Cygni	19 26·3	+27 44	3	7	55·1	34·32	1879·7	β.	A beautiful pair, colours golden yellow and smalt blue.
58.	β Geminorum	7 38·6	+28 18	2	14	274·9	43·00	1877·9	β.	Disc. by Burnham, who also finds the companion double; dist. 1″·4 (1879·2).
59.	α' Capricorni	20 11·9	−12 53	......	14	219 7	44·55	1879·7	β.	α¹ and α² Capricorni (No. 43) form a naked-eye double; Pos. 291°, Dist. 373″·4.
60.	α Canis Min.	7 33·6	+ 5 30	1	14	317·3	44·62	1877·9	β.	Difficult object; just seen steadily by Dawes with 8½-inch refractor.
61.	β Lyræ, A.B.	18 46·0	+33 14	3	7	148·9	45·20	1886·9	T.	There are three other faint and distant components.
62.	α Lyræ	18 33·2	+38 41	1	11	156·1	48·00	1879·7	β.	Good light-test for a 3-inch. There are other more distant companions.
63.	α Cassiopeiæ	0 34·3	+55 56	2	13½	280·2	61·33	1879·7	β.	The 36-inch refractor shows a very faint *comes*; Dist. 17″·5 (Burnham).
64.	α Canis Maj., A.C.	6 40·3	−16 34	1	13	114·9	71·39	1877·5	Hall.	This faint and distant companion to Sirius was disc. by Marth.
65.	α Andromedæ	0 27	+28 29	2	11	271·6	71·60	1878·6	G.	A wide double, visible in a 3-inch, but *comes* very faint.
66.	α Tauri	4 29·6	+16 17	1	12	34·1	114·96	1879·7	β.	Good light-test for a 3-inch. Very faint *comes* Pos. 109°; Dis. 30″·4 (Burnham).

rienced observers, conscious of this advantage, have secured excellent measures in daylight. Mr. Gledhill says:—" Such stars as γ Leonis and γ Virginis are best measured before or very soon after sunset" ('Observatory,' vol. iii. p. 54).

The determination of the angles of position and distance of double stars forms a very important and extensive branch of work in connection with sidereal astronomy. In cases where double stars preserve stationary places relatively to each other, there is clearly no need for frequent re-observation. But in those numerous instances where the two components form a binary system it is desirable to obtain as many measures as possible, so as either to verify the calculated orbit or to furnish the materials for an orbit if one has not been computed before. Dr. Doberck, whose name is well known in these researches, mentioned, in 1882, that ample data for purposes of computation had not been secured for more than thirty or forty binaries out of between five and six hundred such systems that were probably known to exist. Sir W. Herschel, in 1803, estimated the period of revolution of α Geminorum as 342 yrs. 2 mths. and of γ Virginis as 1200 yrs. Orbits* do not appear, however, to have been computed until 1827, when Savery of Paris showed that the companion of ξ Ursæ Majoris was revolving in an ellipse with a period of $58\frac{1}{4}$ years. The accomplished Encke also turned his attention to this work, and adopted a more elaborate method; and many others have pursued the subject with very interesting and valuable results. On pp. 302–305 is a selected list of some of the most noteworthy double and binary stars, arranged according to the distance between the components.

In compiling the above list, I have used some of the latest measures available, as most of these doubles are binary systems, and therefore in motion, so that a few years effect a perceptible difference in the angles of position and distance of the components. Some of the pairs are closing up, others are opening, and thus it happens that a binary star, divided with

---

* Dr. Doberck gives some valuable information with reference to the computation of binary star-orbits in 'The Observatory,' vol. ii. pp. 110 and 140.

great difficulty to-day, may become an easy object some years hence, and *vice versâ*. In fact, as telescopic tests they are constantly varying.

Before leaving this part of the subject it may be interesting to refer individually to a few brilliant examples of double stars.

a Canis Majoris (*Sirius*). A red star according to ancient records, but it is now intensely white. In 1844 Bessel inferred from certain little irregularities in the proper motion of this star that it consisted of a binary system with a period of about half a century*. Peters confirmed this idea in 1851, and it was observationally verified eleven years afterwards. On Jan. 31, 1862, Alvan Clark, jun., while testing a new 18½-inch refractor, discovered a very faint companion 10″ distant. Measures in the few subsequent years proved that the position-angle was decreasing, while the distance showed a slight extension. In 1872 it was about 11″·50, but since then the two stars have been approaching each other, and Mr. Burnham's measures in April 1890 gave the distance as only 4″·19. It is now, therefore, a very difficult object, and only visible in large instruments. In England it is never easy, owing to its southern position, and it has been little observed, but it is satisfactory to note that the large refractors at Washington, Princeton, and Chicago, U.S.A., have been often employed on this object in recent years. Mann gives a period of 51·22 years for this interesting binary, and places the time of periastron-passage as 1890·55. This differs from Gore's orbit, quoted in the table.

β Orionis (*Rigel*). A favourite test-object for small instruments. The companion has been seen with only 1½-inch aperture by experienced observers familiar with the object,

---

* The star *a* Canis Minoris (Procyon) was also inferred to be a binary and to have a similar period. Several close companions appear to have been discovered (Ast. Nach. no. 2080). But Prof. Hall, using the 25·8-inch refractor at Washington, says:—I have never been able to see any of these companions that would stand the test of sliding and changing the eyepiece, turning the micrometer, &c., and am therefore doubtful of their existence. This is an interesting star for the powerful telescopes of the future." It has been surmised that the companion is a non-luminous one, and therefore invisible.

and accustomed to its appearance in larger telescopes. The beginner may, however, esteem himself fortunate if he distinguishes the smaller star with 3 inches of aperture. When he has done this he may afterwards succeed with $2\frac{1}{2}$ inches only, and quite possibly with 2 inches. He can ascertain his ability in this direction by inserting cardboard diaphragms of the diameters referred to in the dew-cap of his telescope. This object is not a binary; the component stars are fixed relatively to each other, and merely form an optical double. The colours are pale yellow and sapphire blue. Burnham thought the smaller star was elongated, as though a very close double, but the 36-inch at Mount Hamilton has disproved the idea.

α *Lyræ* (*Vega*). Another well-known object, and one upon which amateurs are constantly testing their means. The companion star is extremely faint, and small instruments would have no chance with it but for its comparatively wide distance from Vega. Were it much nearer it would be obliterated in the glare. This is a more difficult pair than that of Rigel, though certain lynx-eyed observers have glimpsed the minute star with ridiculously small apertures. It is no mean feat, however, to discern the star with a 3-inch telescope. Webb saw it more easily with a power of 80 than with 144 on a $3\,\tfrac{7}{10}$-inch. There are many other stars in the same field, though more distant than the companion alluded to. With power 60 on my 10-inch reflector, I counted eighteen stars in the field with Vega on Oct. 9, 1889, though the full Moon was shining at the time. Several faint stars have been alleged to exist much closer to Vega than the well-known *comes*; but these have resisted the great American refractors, and it may be safely assumed that they were ghosts produced by a faulty image.

α *Ursæ Minoris* (*Polaris*). This double, from its constant visibility in northern latitudes, from its unvarying brightness, and from the relatively stationary positions of the stars composing it, forms an excellent test for small instruments. But it is a comparatively easy object, and ought to be seen in a 2-inch telescope. With this aperture the primitive efforts of a young observer will probably be disappointing. If possible he should first look at the pair through a 3- or 4-inch, and then he will know exactly what he may expect to see with inferior means. A difficult object is often readily glimpsed in a small telescope after the eye has become acquainted with it in a larger one. Experience of this kind is very requisite, and it is by thus educating the eye that observers are gradually enabled to reach objects which appeared hopelessly beyond them at their first attempts. The companion to Polaris, like that of Rigel and Vega, though situated in nearly the same line of sight is not physically related to the larger star, the contiguity of the objects being accidental. Some dubious observations have been made of *comites* nearer to Polaris than the one to which we have been adverting; but Burnham does not see these, and we are forced to conclude that they have no objective existence.

α *Scorpii* (*Antares*). A fiery-red star, with a rather close, faint companion. This object being in 26° of S. declination is rarely seen with advantage in places with latitudes far north. Atmospheric disturbance usually affects the image in such degree that the smaller star is merged in the contortions of the larger. This pair is, however, interesting from the circumstance that it is frequently liable to occultation by the Moon. A night on which this double star can be distinctly seen may be regarded as an exceptional one in point of definition. It appears to have been discovered nearly half a century ago by Grant and Mitchel.

Variable Stars.—A proportion of the stars exhibit fluctuations in their visible brightness. In most cases, however, the variation is but slight, though there are instances in which the differences are considerable. The fluctuations are periodical in nature and capable of being exactly determined. But the character of the variation and the period are very dissimilar in different stars. Some are of short period, and their variations occupy a few days only ; others, however, are more gradual, and twelve months or more may represent the complete cycle of their changes. These alterations of brightness generally escape the notice of casual observers of the heavens. To them the stars appear as constant in their relative magnitudes as they are in their relative positions. But a close observer of the firmament, who habitually watches and records the comparative lustre of the stars, must soon discover numerous evidences of change. He will remark certain stars which alternately grow bright and faint, and, in fact, display a regular oscillation of brilliancy. In the case of a pair of stars he may possibly notice that the superior lustre is emitted first by one and then by the other. The observation of these variables dates from a period anterior to the invention of the telescope. Nearly three centuries ago Fabricius remarked that o Ceti (Mira) suffered a great diminution of light ; for while it was of the 3rd mag. in Aug. 1596, it became invisible in the following autumn. It was re-observed by Holwarda in 1639, and as he appears to have been the first to estimate its period, some authors, including Argelander, have credited him with the discovery. The star has a period of about 331·3 days. Its variations are somewhat erratic, for at max. it is sometimes only 4th mag., while at others it is as bright as 2nd mag., and its min. are equally inconsistent.

β Persei (Algol) is another and perhaps the best known of all the variable stars. Its changes are very rapid, for it passes through its various gradations of brilliancy in less than three days. It was first noticed by Montanari in 1669, though it was left for Goodricke in 1782 to ascertain its period. The normal mag. of the star is 2·2, and it only shows distinct variation during the five hours which precede and follow a minimum, when it declines to 3·7 mag. Its period is shortening, for in 1782 it was 2^d 20^h 48^m 59^s·4, in 1842, 2^d 40^h 48^m 55^s·2, and at present Chandler finds it 2^d 20^h 48^m 51^s. As to the causes which contribute to these variations, they are invested in mystery. It has been conjectured that dark spots on the surfaces of the stars may, by the effects of rotation, introduce the observed alternations. Another surmise is that the temporary diminutions of lustre are to be ascribed to the interposition of dark satellites, and this theory seems tenable in regard to stars of the Algol type. It is satisfactory to note that a large amount of systematic work is being done in this important and delicate branch of research. Such stars as are subject to variation have been classed as follows :—
1. Temporary or new stars ; 2. Stars having long and pretty regular variation ; 3. Stars irregularly variable ; 4. Stars

List of Variable Stars.

Name of Star.	Position, 1890.		Mags.	Period.	Observer.
	R.A.	Dec.			
	h m	° ′			
μ Cephei	0 52·5	+81 17	7·2, 9·4	2d 11h 50m	Ceraski, 1880.
o Ceti	2 13·8	− 3 29	2, 0	331⅓ days	Fabricius, 1596.
β Persei	3 1·0	+40 32	2·2, 3·7	2d 20h 49m	Montanari, 1669.
λ Tauri	3 54·6	+12 11	3·4, 4·2	3d 22h 52m	Baxendell, 1848.
U Orionis	5 49·3	+20 9	6, 12½	......	Gore, 1885.
η Geminorum	6 8·2	+22 32	3·2, 4·2	135–151 days	Schmidt, 1865.
ζ Geminorum	6 57·6	+20 44	3·7, 4·5	10d 3h 43m	Schmidt, 1847.
L₂ Puppis	7 10·2	−44 28	3·5, 6·3	136 days	Gould, 1872.
R Canis Maj.	7 14·5	−16 11	6·2, 6·8	1d 3h 16m	Sawyer, 1887.
U Geminorum	7 48·6	+22 18	9, 14	71–126 days	Hind, 1855.
S Cancri	8 37·7	+19 26	8·2, 11·7	9d 11h 38m	Hind, 1848.
η Argûs	10 40·8	−59 6	1, 6	46 or 67 yrs.?	Burchell, 1827.
R Hydrae	13 23·7	−22 43	4, 10	436 days	Maraldi, 1704.
δ Libræ	14 55·1	− 8 5	4·9, 6·1	2d 7h 51m	Schmidt, 1859.
U Coronæ	15 13·7	+32 3	7·6, 8·8	3d 10h 51m	Winnecke, 1869.
α Herculis	17 9·6	+14 31	3·1, 3·9	88d 12h (irreg.)	W. Herschel, 1795.
U Ophiuchi	17 11·0	+ 1 20	6, 6·7	0d 20h 8m	Sawyer, 1881.
β Lyræ	18 46·0	+33 14	3·5, 4·5	12d 21h 47m	Goodricke, 1784.
χ Cygni	19 46·3	+32 38	+6·5, 13	406 days	Kirch, 1686.
η Aquilæ	19 46·9	+ 0 44	3·6, 4·7	7d 4h 14m	Pigott, 1784.
Y Cygni	20 47·7	+34 15	7·1, 7·9	1d 11h 57m	Chandler, 1886.
μ Cephei	21 40·1	+58 16	3·6, 4·8	432 days?	Hind, 1848.
δ Cephei	22 25·1	+57 51	3·7, 4·8	5d 8h 48m	Goodricke, 1784.

varying in short periods ; 5. Stars of the type of Algol, which are liable to temporary diminutions of lustre. On the preceding page is a list of the most noteworthy variable stars.

New or Temporary Stars.—These stars (sometimes classed with variable stars) furnish us with rare instances of vast physical changes occurring among sidereal objects, usually so steadfast and endurable. The alternating lustre of certain variable stars represents phenomena of regular recurrence, and is probably to be explained by simple causes ; but the sudden outbursts and rapid decline of temporary stars are facts of a more startling character, and need a more exceptional explanation. The first of these objects recorded in history appears to have been seen in Scorpio 134 years before the Christian era, and it suggested to Hipparchus of Rhodes the idea of forming a catalogue of stars, so that in future ages observers might have the means of recognizing new stars or any other changes in the configuration of the heavens. Hipparchus completed his catalogue in 128 B.C.; it contained 1025 stars, and forms one of the most valuable memorials we possess of the labours of the ancient astronomers. Another temporary star is said to have appeared in 130 A.D., but this and several other objects of presumably similar character noticed in later years may just possibly have been comets, and considerable doubt hangs over the descriptions. It will therefore be safest to confine our remarks to more modern and better attested instances of these phenomena*:—

1572, November 11.—The famous star of Tycho Brahe. He thus described his first view of it :—" One evening when I was considering, as usual, the celestial vault, the aspect of which is so familiar to me, I perceived with indescribable astonishment a bright star of extraordinary magnitude near the zenith in the constellation of Cassiopeia." He adds :— " The new star was destitute of a tail, or of any appearance of nebulosity; it resembled the other stars in all respects, only that it twinkled even more than stars of the first magnitude. In brightness it surpassed Sirius, *a* Lyræ, or Jupiter. It could

---

* It is remarkable that nearly all the temporary stars have appeared in the region of the Milky Way.

be compared in this respect only to Venus when she is nearest the earth (when a fourth part of her illuminated surface is turned towards us). Persons who were gifted with good sight could distinguish the star in the daytime, even at noon, when the sky was clear." This brilliant NOVA began to fade early in Dec. 1572, and in April and May 1573 it resembled a star of the 2nd mag., in July and Aug. one of the 3rd mag., and in Oct. and Nov. one of the 4th mag. In March 1574 the star completely disappeared (to the naked eye), after a visibility of about 17 months. It exhibited some curious variations of colour. It was white when most brilliant; it then became yellow, and afterwards red, so that its hue in the early part of 1573 was similar to that of Mars. But in May it again became white, and continued so until it ceased to be visible. The position of this star (for 1890) is R.A. 0^h 18^m 41^s, Dec. $+63°$ $32'\cdot2$. It was supposed to be a reapparition of the brilliant stars which shone between Cepheus and Cassiopeia in 945 and 1264, and to have possibly been associated with the " Star of Bethlehem ;" but there is no reliable evidence on which this view can be supported, as the alleged " stars " of 945 and 1264 were undoubtedly comets misdescribed in old records. Cornelius Gemma is reputed to have seen the celebrated star of 1572 a few days before Tycho Brahe, viz., on November 8, 1572.

1604, October 10.—Discovered by Brunowski, who announced it to Kepler. It was brighter than a star of the 1st mag., also than Mars, Jupiter, or Saturn, which were not far distant at the time. It did not begin to fade immediately; for a month after its discovery it was still brighter than Jupiter, and of a white lustre. At the middle of November it surpassed Antares, but was inferior to Arcturus. In April 1605 it had fallen to the 3rd mag., and went on decreasing until in October it could scarcely be seen with the naked eye owing to the twilight resulting from its proximity to the Sun. In March 1606 it was invisible. The position of this object was about midway between ξ and 58 Ophiuchi, or at R.A. 17^h 24^m, Dec. $-21°$ $20'$ (1890).

1670, June 20.—Discovered by the Carthusian Monk Anthelme in R.A. 19^h 43^m 3^s, Dec. $+27°$ $3'$ (1890), a few degrees east of β Cygni. It was of the 3rd mag., and continued in view, with constantly fluctuating brightness, for nearly two years. At the end of March 1672 it was 6th mag., and has never reappeared. Hind found a small, hazy, and ill-defined star in the same place, but this is probably not the same as Anthelme's star of 1670.

1848, April 28.—During the long interval of 178 years separating 1670 from 1848 not a single new star appears to have revealed itself. Observers had multiplied, astronomical instruments had been much improved, star-catalogues were plentiful, and yet the sidereal heavens gave no intimation of a stellar outburst. No better proof than this could be afforded as to the great rarity of temporary stars. At length, in the spring of 1848, the spell was broken, and Mr. Hind announced that a new star of a reddish-yellow colour had appeared in Ophiuchus, R.A. 16^h 53^m 20^s, Dec. $-12°$ $43'$ (1890). He expressed himself as certain that no star brighter than the 9th mag. had been there previous to April 5. After rising to the 4th mag. it soon faded, and in 1851 could only be observed in large instruments. In 1875 it was still visible as a very minute star.

1860, May 21.—M. Auwers, of Konigsberg, noticed a star of the 7th mag. near the centre of the bright resolvable nebula (M. 80), lying between α and β Scorpii, R.A. 16^h 10^m 29^s, Dec. $-22°$ $42'$ (1890). On May 18 the star was not there, and it disappeared altogether in three weeks. It was independently seen by Pogson on May 28, and to him it seemed that "the nebula had been *replaced* by a star, so entirely were its dim rays overpowered by the concentrated blaze in their midst."

1866, May 12.—Discovered by Birmingham at Tuam. It was of the 2nd mag., and situated in Corona, R.A. 15^h 54^m 54^s, Dec. $+26°$ $14'$ (1890). The outburst must have been very sudden, as Schmidt, at Athens, was observing this region three hours before the new star was detected, and is certain it was then fainter than the 4th mag. The star was found to be identical with one on Argelander's charts estimated as $9\frac{1}{2}$ mag. It faded from the 2nd to the 6th mag. by May 20, and was thereafter invisible to the naked eye.

1876, Nov. 24.—A yellow star of the 3rd mag. was seen by the ever vigilant Schmidt at Athens near ρ Cygni, and where no such star existed on Nov. 20. The position of the object was R.A. 21ʰ 37ᵐ 23ˢ, Dec. +42° 20' (1890). It soon grew fainter, so that on Dec. 13 it was of the 6th mag. and devoid of colour. In the spectroscope it presented much the same lines as Birmingham's star of May 1866. In addition to the continuous spectrum it showed bright lines of hydrogen.

1885, August 31.—Dr. Hartwig announced the appearance of a star-like nucleus in the great nebula (M. 31) of Andromeda, R.A. 0ʰ 36ᵐ 43ˢ, Dec. +40° 40' (1890). Other observers soon corroborated the discovery. The star appears to have been first seen on Aug. 19 ; it was not visible on the preceding night. On Sept. 1 its mag. was 6·5, on Sept. 2, 7·3, on Sept. 3, 7·2, Sept. 4, 8·0, Sept. 18, 9·2, &c. On Feb. 7, 1886, it had dwindled down to the 16th mag., according to an estimate made by Prof. Hall with the great Washington refractor. The spectrum was continuous, and Proctor and Gore considered " that the evidence of the spectroscope showed that the new star was situated *in* the nebula."

The phenomena presented by the temporary stars alluded to are so different to those of ordinary variables that it is very questionable whether they ought to be classed together. Our knowledge of the former would no doubt progress more rapidly were they specially looked for and more instances discovered. Those who have acquired a familiar acquaintance with the naked-eye stars should examine them as often as possible with this end in view. Some of these objects lose light so quickly that unless they are caught near the maximum they are likely to escape altogether, and this shows the necessity of being constantly on the alert for their appearance. I have frequently, while watching for meteors, reviewed the different constellations in the hope of picking up a new object, but have never succeeded in doing so.

Star Colours form another interesting department of sidereal astronomy. It is obviously desirable to record the hues presented, not only by double stars and binary systems, but by isolated stars also, as changes of tint have been strongly suspected. Cicero, Seneca, Ptolemy, and others speak of

Sirius as a red star, whereas it is now an intense white; and if we rely on ancient descriptions similar changes appear to have affected some other prominent stars. But the old records cannot be implicitly trusted, owing to the errors of transcribers and translators ; and Mr. Lynn ('Observatory,' vol. ix. p. 104) quotes facts tending to disprove the idea that Sirius was formerly a red star. In the majority of cases double stars are of the same colour, but there are many pairs in which the complementary colours are very decided. Chambers remarks that the brighter star is usually of a ruddy or orange hue, and the smaller one blue or green. " Single stars of a fiery red or deep orange are not uncommon, but isolated blue or green stars are very rare. Amongst conspicuous stars β Libræ (green) appears to be the only instance." As an example of fiery-red stars Antares may be mentioned ; Aldebaran is deep reddish orange, and Betelgeuse reddish orange. Amongst the more prominent stars Capella, Rigel, and Procyon may be mentioned as showing a bluish tinge, Altair and Vega are greenish, Arcturus is yellow, while Sirius, Deneb, Polaris, Fomalhaut, and Regulus are white. Mr. Birmingham published a catalogue of " The Red Stars " in the ' Transactions of the Royal Irish Academy, for August 1877, and Mr. Chambers has a *working-catalogue* of 719 such objects in the ' Monthly Notices,' vol. xlvii. pp. 348–387. The region of Cygnus appears to be especially prolific in red stars, and many of these objects are variable. In a paper read at a recent meeting of the Astronomical Society of the Pacific Mr. Pierson stated that in binary systems where the stars are of equal magnitude the colours are invariably the same, while those differing in magnitude differ also in colour and the larger star is always nearer the red end of the spectrum than its secondary. In the estimation of star-colours reflecting-telescopes are very reliable owing to their perfect achromatism.

Groups of Stars.—Great dissimilarity is apparent in the clustering of stars. The heavens furnish us with all gradations—from the loose, open groups like that in Coma Berenices, in the Pleiades, or in Præsepe, to the compact globular clusters, in which some thousands of stars are so densely congregated that considerable optical power is required to disintegrate them. Some, it is true, yield more easily than others. The great cluster (Messier 13) in Hercules readily displays the swarms of stars of which it is composed ; but others are so difficult that it is only in the largest instruments they are resolved into star-dust. Further references to these wonderful objects will be made in the next chapter, and some of the principal examples described ; our purpose here is to allude to a few of the more scattered groups, and to some noteworthy instances of multiple stars.

Coma Berenices. A naked-eye cluster, consisting of many stars, chiefly from the 5th to 6th mags. A telescope adds a number of smaller stars. Nebulæ may be often swept up hereabouts, as it is not far north of the rich nebulous region of Virgo.

The Pleiades. Six stars are usually distinguished by the naked eye, and a seventh is occasionally remarked. Möstlin (the instructor of Kepler) counted fourteen, Miss Airy has drawn twelve, and Carrington, like Möstlin, saw fourteen. In 1877 I distinctly made out fourteen stars in this group. The telescope reveals a considerable number of small stars and Tempel's large nebula near Merope. Kepler saw thirty-two stars with a telescope, and Hooke seventy-eight ; but Wolf, at Paris, after three years of unremitting labour with a 4-foot reflector, catalogued 671 stars in the group. A photograph, however, with a 12-inch refractor showed 1421 stars ; and a more recent negative includes no less than 2326. There is an interesting little triangle close to the brightest star, Alcyone ; and several of the leading stars are involved in nebulosity, discovered by means of photography.

Præsepe. A fine group of small stars, divisible by the unaided eye on a clear night. Chambers says the components are not visible without a telescope ; while Webb notes that the group is just resolvable by the naked eye. Thirty-six stars were glimpsed with Galilei's telescope ; but modern instruments show many more. Marth, using Lassell's 4-foot reflector at Malta, discovered several faint nebulæ and nebulous stars in this cluster.

χ *Persei.* Perceptible to the eye as a patch of hazy material lying between the constellations Cassiopeia and Persei. In a telescope it forms a double cluster, and is one of the richest and most beautiful objects that the sky affords. The tyro who first beholds it is astonished at the marvellous profusion of stars. It can be fairly well seen in a good field-glass, but its chief beauties only come out in a telescope, and the larger the aperture the more striking will they appear. It is on groups of this character that the advantage of large instruments is fully realized. The power should be very low, so that the whole of the two clusters may be seen in the field. An eyepiece of 40, field 65', on my 10-inch reflector, presents this object in its most imposing form.

κ Crucis. Sir J. Herschel's observations at the Cape have made this object familiar to northern observers. It is composed of more than 100 stars, from the 7th mag. downwards; and some of the brighter ones are highly coloured, so that the general effect is greatly enhanced and fully justifies Herschel's statement that the group may be likened to " a superb piece of fancy jewelry."

ζ Ursæ Majoris (Mizar). This group is interesting both as a naked-eye and as a telescopic object. There is a 5th mag. star, named Alcor, about $11\frac{1}{2}'$ distant from Mizar, and the former was considered a good test-object for unaided vision by the Arabian astronomers. But the star has probably brightened; for it can now be easily seen, and certainly offers no criterion of good vision. Mizar is a fine telescopic double, the companion being 4th mag. and distant $14\frac{1}{2}''$. Any small telescope will show it, and there is another 8th mag. star very near.

σ Orionis. This appears as a double-quadruple star, with several others in the same field. A 3-inch will reveal most of them, though some of the fainter stars in the group will be beyond its reach.

θ Orionis. In the midst of the great nebula of Orion there is a tolerably conspicuous quadruple star, the components of which form a trapezium. This is visible in a 2-inch refractor. In 1826 Struve discovered a fifth star, and in 1830 Sir J. Herschel found a sixth; these were both situated a little outside the trapezium. All these stars have been seen in a 3-inch telescope.   The great 36-inch equatoreal at Mount Hamilton has added several others ; one was detected by Alvan G. Clark (the maker of the object-glass) and another by Barnard.   These were excessively minute, and placed within the trapezium.   Barnard* has also glimpsed an extremely minute double star exterior to the trapezium, and forming a triangle with the stars A and C on the following diagram :—

Fig. 63.

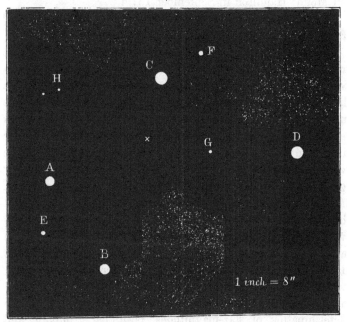

The Trapezium in Orion, as seen with the 36-inch refractor.

Several observers, including Huggins, Salter, and others, had previously drawn faint stars in the interior of the trapezium ; but these could not be seen by Hall and Burnham in the large refractors at Washington and Chicago, and were thus proved to have no real existence.   The new stars observed in the 36-inch telescope are only just within the limits of its capacity, and therefore cannot be identical with

* This expert comet-finder would appear to have more acute, sensitive vision on faint stars than Burnham (see 'Monthly Notices,' vol. xlix. p. 354).

stars alleged to have been previously seen in small instruments. The fifth and sixth stars in the trapezium have been
supposed to be variable, and not without reason ; possibly the
others are equally liable to change, but this is only conjecture.
Sir J. Herschel says that to perceive the fifth and sixth stars
" is one of the severest tests that can be applied to a telescope " ('Outlines,' 11th edit. p. 610); yet Burnham saw
them both readily in a 6-inch a few minutes before sunrise
on Mount Hamilton in September 1879.

β and ϵ Lyræ also form multiple groups, which will well
repay observation either with large or small telescopes.

Further Observations.—Anyone who attempts to indicate
with tolerable fulness the methods and requirements of observation in the stellar department of astronomy will find a
heavy task lies before him ; and it is one to which he will be
unable to do justice in a small space, owing to the variety of
matters to be referred to and the necessity of being particular
in regard to each one. In what follows I shall merely make
very brief allusions, as it is hoped the description already
given of past work will be a sufficient guide for the future.
Moreover, those who take up a special branch of inquiry will
hardly rest content with the meagre information usually conveyed in a general work on astronomy, but will consult those
authorities who deal more exclusively with that branch.
Double and binary stars may be said to form one department,
variable and temporary stars another, the colours of stars
a third, while many others may be signified—such as the
determination of star-magnitudes, positions, grouping, and
distances ; also the proper motions and number of stars,
besides photographic and spectroscopic work,—each and all
of which comprise a field of useful and extensive inquiry.
The amateur will of course choose his own sphere of labour,
consistently with his inclination and the character of his
appliances. In connection with double stars, valuable work
yet remains to be done, though the Herschels and the Struves
gathered in the bulk of the harvest and Burnham has gleaned
much that was left. With regard to bright stars, it may be
assumed that very few, if any, close companions, visible in
moderately small glasses, now await discovery, unless, indeed,

in cases where the star forms part of a binary system of long period, and the epoch of periastron has fallen in recent years. But with telescopic stars there must be many interesting doubles, some of them binaries, still unknown. These should be swept up and submitted to measurement. A great desideratum in this branch is a new general catalogue of double stars; for such a work would greatly facilitate reference, and save the trouble of searching through different lists in order to identify an object. Burnham has given some practical hints on double-star work in the 'Sidereal Messenger,' and his remarks are reproduced in that excellent work 'Astronomy for Amateurs.'

As to variable stars, some of these permit of naked-eye estimation, others need a field-glass, and there are some which require to be followed in a good telescope. The observer who enters this department may either desire to find new objects or to obtain further data with regard to old ones. If the former, he cannot do better than watch some of the suspected variables in Gore's Catalogue of 736 objects, published by the Royal Irish Academy. Whether suspected or known variables are put under surveillance, the plan of comparison will be the same. Several stars near the variable in position, and nearly equal in light, should be compared with it, and the differences in lustre, in tenths of a magnitude, recorded as frequently as possible. The extent and period of the variation will become manifest by a discussion of the results. The comparison-stars should of course be constant in light, and, if naked-eye stars, they may be selected from the *Uranometria Nova Oxoniensis* or 'Harvard Photometry.' If telescopic stars are required, then recourse must be had to comprehensive charts such as Argelander's *Durchmusterung*, which includes stars up to $9\frac{1}{2}$ mag. Variable stars of the Algol type are especially likely to escape recognition, as they retain a normal brilliancy except during the few hours near the time of a minimum.

As to star-colours, it must be admitted that our knowledge is in an unsatisfactory condition. The results of past observation show discordances which are difficult to account for. When, however, all the circumstances are considered, we

Y

need feel no surprise at this want of unanimity. In certain cases it is probable that actual and periodical changes occur in the colours of stars, though absolute proof is still required. Atmospheric variations unquestionably affect the tints of stars, and some alterations depend upon altitude, for a celestial object seen through the dense lower air-strata near the horizon will hardly preserve the same apparent hues when on the meridian. Telescopes are also liable to induce false impressions of colour, and especially by the employment of different eyepieces not equally achromatic. And the observer's judgment is sometimes at fault through physiological influences, or he may have a systematic preference for certain hues which little impress another observer. Those engaged in this branch feel the want of a reliable and ready means of comparison, and several have been tried ; but there are objections to their use, and it seems that the best objects are furnished by the stars themselves. Let the observer study the colours of well-known stars, and familiarize his eye with the distinctions in various cases (also with the differences due to meteorological effects &c.) ; he will then gradually acquire confidence, and may use these objects as standards. The difficulty will be that they cannot be directly compared, in the same field, with other stars ; but relative differences may be noted by turning the telescope from one object to the other. This will be better than forming estimates on the basis of an artificial method, which will sure to be troublesome to arrange, and probably erroneous in practice. In some stars the colour is so curious as to be attributed with difficulty, and with regard to faint stars colour-estimates are often unreliable ; so that it is not desirable to go below the 9th mag. unless a very large instrument is employed.

The necessity of being constantly on the look-out for temporary stars has been already mentioned. There is also the need for further observations of such of these objects as still exist. They are, however, very minute, and the observer will have to be careful as to their identity. Though no great revival in brilliancy is to be expected, these objects exhibited some singular fluctuations during their decline, and it is important to keep them under view as long as possible.

Many other departments of sidereal work are best left to the professional astronomer. The derivation of accurate star-places, proper motions, distances, &c. requires instruments of great refinement and trained hands to use them. Researches such as these do not come within the scope of ordinary amateurs. But a vast field is open to them in respect to double and variable stars ; and the physical relations of many of the former greatly intensify the interest in this branch, and make it necessary to secure frequent observations.

CHAPTER XVII.

NEBULÆ AND CLUSTERS OF STARS.

Distinction.—Large number of Nebulæ and Clusters visible.—Varieties of form and grouping.—Distribution.—Early Observations.—Variable Nebulæ.—Nebulous Stars.—The Magellanic Clouds.—Double Nebulæ.—Real dimensions of Nebulæ and Clusters.—Round Nebulæ and Clusters.—Description of Objects.—Further Observations required.—Lists of selected Objects.

Distinction.—These objects, though classed together in catalogues, offer some great distinctions which the observer will not be long in recognizing. It was thought at one period that all nebulæ were resolvable into stars*, and that their nebulous aspect was merely due to the confused light of remote star-clusters. But modern telescopes, backed up by the unequivocal testimony of the spectroscope, have shown that purely nebulous matter really exists in space. The largest instruments cannot resolve it into stars, and it yields a gaseous spectrum. The conjecture has been thrown out that it may be considered as the unformed material of which suns and planets are made.

Large Number visible.—D'Arrest once said that nebulæ are so numerous as to be infinite, and his opinion is supported by the rapid increase in the number known. Let us make a comparison. Messier inserted in the *Connaissances des Temps* for 1783 and 1784 (published in 1781) a catalogue containing 103 nebulæ and star-clusters. Of these 68 were new. In 1888 a new edition of Sir J. Herschel's catalogue

---

* Sir W. Herschel at first entertained this view, finding that with every increase of telescopic power more nebulæ were resolved. But in 1791 he said, "perhaps it has been too hastily surmised that all milky nebulosity is owing to starlight only." Lacaille had remarked in 1755 that "it is not certain the whiteness of parts of the Milky Way is caused by clusters of stars more closely packed together than in other parts of the heavens.

of 1864 (revised and extended by Dreyer) was printed by the
Royal Astronomical Society, and this includes 7840 objects ! *
The labours of the Herschels, of Lord Rosse, D'Arrest, Marth,
Tempel, Stephan, and Swift have vastly augmented our
knowledge in this branch since the time of Messier.

Varieties of Form and Grouping.—A telescope reveals all
grades of condensation in stellar groups. Some consist of
rather bright, scattered stars, and are easily resolved. Others
contain more stars, but they are smaller, and greater power
is required to show them. Others again are condensed into
globular clusters needing high powers and good instruments
to disconnect the mass of stars composing them. Some are
faint, and the stars so minute that they are only to be distinguished from nebulæ in the finest telescopes. As to the
nebulæ properly so called, they exist in all forms. They may
be either round, elliptical, or in the form of a streak. Some
are highly condensed in their centres, others present well-
defined circular disks like planets, and a small proportion are
in the form of rings†. Many peculiarities of detail have been
remarked, and a curious and complicated spiral structure has
been discovered in certain prominent nebulæ. One of these
has been termed the "Whirlpool" Nebula from its singular
convolution of form. Other objects have received distinctive
appellations agreeably to their appearance. Thus, there is the
"Dumb-bell" Nebula, the "Crab" Nebula, the "Horseshoe"
Nebula, &c. Lord Rosse's 6-foot reflector is in a large degree
responsible for the particular knowledge we possess of many
of these objects. The large mirror commands a grasp of light
which renders it very effective on forms of this character.
An instrument of small diameter is quite inadequate to deal
with them. They can be seen, it is true, and the general
shape recognized in the most conspicuous examples, but their
details of structure are reserved for the greater capacity of
large apertures.

* This is exclusive of 47 new nebulæ discovered by Prof. Safford, which
form the appendix to the catalogue.

† Chambers says only four examples are known, but this is erroneous,
as Lord Rosse's telescope has added five ring-nebulæ to the four previously catalogued.

Distribution.—With regard to distribution these objects exhibit the utmost irregularity, for in certain regions of the heavens they are found to be very plentiful, while in others they are singularly rare. Thus, in Virgo, Coma Berenices, Leo, and Ursa Major large numbers of nebulæ abound, while in Hercules, Draco, Cepheus, Perseus, Taurus, Auriga, &c., very few are encountered. Taking the 7840 objects in the New General Catalogue of 1888 it will be found that their distribution in hours of Right Ascension is as follows:—

R.A.	Nebulæ.	R.A.	Nebulæ.
0 H. . . .	387	XII H. . . .	858
I	428	XIII	504
II	398	XIV	375
III	300	XV	212
IV	276	XVI	230
V	375	XVII	259
VI	171	XVIII	203
VII	196	XIX	117
VIII	230	XX	153
IX	362	XXI	188
X	404	XXII	275
XI	585	XXIII	354

The maximum is therefore reached at XII hours, while the minimum is shown at XIX h. There is a secondary max. at I h., and a secondary min. at VI h.

Early Observations.—The nebula in Andromeda appears to have been the one first discovered, for the distinguished Persian astronomer Al-Sûfi (who died in 986 A.D.) was undoubtedly acquainted with it. The nebula is figured upon a Dutch map of the stars nearly 400 years old. In 1612 Simon Marius redetected this object, and appropriately likened its appearance to that of a " candle shining through a piece of horn." In 1618 the nebula in Orion was certainly known, for Cysatus of Lucerne compared it with the head of the fine comet visible in December of that year. Huygens alighted upon the same object in 1656, and appears to have been unconscious of its prior discovery. Only six " nebulæ or lucid spots " were known in 1716, and enumerated by Halley in the ' Phil.

Trans.' vol. xxix. These included those of Andromeda and Orion. A third was situated in the space between the bow and head of Sagittarius. This is M. 22, and consists of a bright globular cluster of stars. The fourth was the fine star-group involving ω Centauri, which Halley himself found in 1677. The fifth was another fine group in the right foot of Antinous. This is M. 11, and was discovered by Kirch in 1681. The sixth was the magnificent globular cluster (M. 13) in Hercules, discovered by Halley in 1714.

In 1735 the Rev. W. Derham published a list of 16 of these objects, and in 1761 Lacaille summarized 42 nebulæ and star-clusters which he had observed in the southern sky. This was followed by Messier's tables of 45 nebulæ &c. in 1771, and of 103 in 1781 *. But these contributions, important though they severally were, sunk into insignificance beside the splendid results obtained by Sir W. Herschel, who during his prolonged and systematic sweeps of the heavens picked up no less than 2500 new nebulæ and clusters which he formed into three catalogues printed in the ' Phil. Trans.' as follows :—1786, 1000 objects, 1789, 1000 ditto, 1802, 500 ditto.

Variable Nebulæ.—It is in the highest degree probable that changes occur in the physical appearances of certain nebulæ, though the opinion is not perhaps supported by a sufficient number of instances. Until Sir W. Herschel began his review of the heavens very few nebulæ were known, and the information possessed about them was very incomplete. The early records, obtained with small and inferior telescopes, scarcely admit of comparison with recent observations, for in matters of detail little agreement will be found; and this proceeds certainly not so much from real changes in the objects as from differences due to the variety of instruments employed, to atmospheric vagaries, and to " personal equation." Bullialdus and Kirch in 1667 and 1676 and Le Gentil in 1759 supposed that remarkable changes were operating in the great elliptical nebula of Andromeda. But G. P. Bond fully

---

* Some of the nebulæ in Messier's list were discovered by Mechain at Paris, who, like Messier, earned celebrity by his cometary discoveries. He was born at Laon in 1744, and died at Valencia in 1805.

investigated the evidence, and concluded that the variability of the object was by no means proved. Some observers have represented the nucleus as stellar, while others have drawn it as a gradual condensation, and Dr. Copeland has shown that different magnifying powers alter the aspect of the nucleus, "the lower powers making it more star-like, the higher ones more soft-looking and extensive."

Mairan and others entertained the view that the large irregular nebula in Orion was subject to change. This object received much attention from Sir W. Herschel, and he concluded that it underwent great alteration between 1774 and 1811. D'Arrest, from his own researches and a discussion of other results, expressed himself in 1872 that "the observed changes in this vast mass of gas seem exclusively to turn out to be temporary fluctuations of brightness." Prof. Holden has arrived at a similar conclusion, and says :—" The figure of the nebula has remained the same from 1758 till now (if we except a change in its apex about 1770, which seems quite possible) ; but in the brightness of its parts undoubted variations have taken place, and such changes are still going on"* ('Monograph of the Nebula in Orion,' p. 225).

Hind discovered a faint nebula, with a diameter of about 1', on Oct. 11, 1852. It was situated in Taurus, the position being R.A. 4^h 15^m 33^s, Dec. $+19°$ $15'·6$ (for 1890), or about $2°$ W. of the star ϵ Tauri (mag. $3·7$). D'Arrest, on Oct. 3, 1861, searched for this object, but found it had quite disappeared ! A small round nebula was seen in 1868, about 4' preceding Hind's, but this resisted some later attempts at observation. In Oct. 1890, Burnham and Barnard, with the 36-inch refractor of the Lick Observatory, saw two nebulæ here, one a very small, condensed nebula, with a stellar nucleus, and the other an exceedingly faint nebulosity about 45" in diameter (see 'Monthly Notices,' vol. li. pp. 94, 95).

The nebula surrounding the star η Argûs has been suspected of variation, particularly by Abbott, of Hobart Town, Tasmania. Vols. xxv., xxx., and xxxi. of the 'Monthly Notices' contain many references to, and figures of, this

* O. Struve had expressed views identical with these in 1857 (see 'Monthly Notices,' vol. xvii. p. 230).

interesting object. But the alleged changes have not been substantiated, and there seems no reason to doubt that they were purely imaginary.

The trifid nebula in Sagittarius (M. 20) is supposed by Prof. Holden to have altered its position with reference to a triple star now situated in the S. following part of the nebula. Sir J. Herschel, more than half a century ago, had described this star as placed in the middle of the vacuity by which the nebula is divided. Dreyer, however, points out that the drawings of this object differ in many details, and that, though changes of brightness may have taken place, it is difficult to understand that the nebula should move so as to envelop the star in about 1835, "after which no sensible change occurred again, so far as published observations go."

The nebula (M. 17) just N. of the bow of Sagittarius was also inferred by Holden to have shifted its place relatively to the small stars figured by Lassell in this object; but Dreyer adduces facts which controvert this assumption. (See 'Monthly Notices,' vol. xlvii. pp. 412–420, where much valuable information will be found as to supposed variable nebulæ.)

On Oct. 19, 1859, Tempel discovered a faint, large nebulosity attached to the star Merope, one of the Pleiades, and at first mistook it for a diffused comet. Its position is R.A. 3^h 39^m·6, Dec. +23° 26′ (1890). An impression soon gained ground that this object was variable; for while Schmidt, Chacornac, Peters, and others saw it with small instruments, it could not be discerned by D'Arrest and Schjellerup with the large refractor at Copenhagen. Swift saw the nebula easily in 1874 with a 4½-inch refractor, and has observed it with the aperture contracted to 2 inches. Backhouse re-observed it in 1882 with a 4¼-inch refractor. Yet in March 1881 Hough and Burnham sent a paper to the Royal Astronomical Society with an endeavour to prove that the nebula did not exist! They had frequently searched for it during the preceding winter, but not a vestige of the object could be seen in the 18½-inch refractor at Chicago, and they regarded the supposed nebula as due to the glow proceeding from Merope and neighbouring stars. But photography has entirely refuted this negative evidence, and has shown, not only Tempel's nebula, but others involving the stars Maia, Alcyone, and Electra belonging to this cluster. As to the alleged variations in the Merope nebula, there is every reason to suppose these were not real.

Proper motion has been suggested in regard to a very small, faint nebula (N.G.C. 3236) a few degrees following *a* Leonis. But Dreyer has disproved this by showing that there was no proper motion between 1865 and 1887, whence "it may be safely inferred that there has been none since 1830, unless we are to believe, in this and similar cases, that nebulæ in the good old days moved about as they liked, but have been on their good behaviour since 1861."

Nebulous Stars.—This name was applied by Hipparchus and other ancient observers to the clusters of stars which, to the naked eye, appear as patches of nebulous light. Sir W. Herschel, in 1791, showed this designation to be incorrect, and used it in connection with stars actually involved in nebulosity. In sweeping the heavens he met with several instances of this kind. Thus, 3° E.S.E. of ζ Persei he found a star of the 9th mag. surrounded by a nebula 3′ in diameter. He picked up another close to the star 63 Geminorum. This is a remarkable object—a star of the 9th mag. surrounded by two dark and two bright rings. On Feb. 3, 1864, Lord Rosse's telescope showed an opening in the outer bright ring, and the latter seemed connected with the inner bright ring; so that the object presented the aspect of a spiral nebula with a star in the centre. The diameter of the whole nebulosity is 45″. Key observed this object with an 18-inch reflector in 1868, and described it as symmetrical—a central star, with intervening dark and bright rings. He found a power of 510 the best, for, "like the annular nebula in Lyra, it bears magnifying wonderfully well." Since Herschel's time many nebulous stars have been discovered. There is one of about 6th mag. in R.A. 8^h $6^m·1$, Dec. $-12°$ 36′. The nebulosity round the star fades away gradually, and its extreme diameter is 157″. There is a 7th mag. star at R.A. 21^h 0^m 14^s, Dec. $+67°$ 44′ involved in a very large, faint nebulosity. This is a striking object, and I have frequently picked it up while comet-seeking. The star has such a foggy, veiled appearance that on first remarking it the observer thinks his lenses are dewed, but on viewing neighbouring stars he sees them sharp and clear on the dark sky, and the contrast is very pronounced. The nebulous star is much isolated, though in a part of the sky where small stars abound. This is one of Herschel's discoveries and No. 7023 of the N. G. C.; Dreyer says he has seen the nebulosity particularly distinct north and south of the star. In some cases a double star is involved in nebulosity, and there are instances in which two double stars are placed within an elliptical nebula.

The Magellanic Clouds *.—These are marked as Nubecula Major and Nubecula Minor on celestial globes and charts. They form two extensive aggregations of nebulæ and star-clusters, and are readily visible to the naked eye in or near Hydrus, and not far from the south pole of the heavens. They may be likened to detached patches of the Milky Way. Sir J. Herschel says the Nubecula Major is situated between the meridians of 4^h 40^m and 6^h and the parallels of 66° and 72° of S. declination, and extends over a space of some 42 square degrees. The Nubecula Minor lies between 0^h 28^m and 1^h 15^m and 72° and 75° of S. declination, and spreads over about 10 square degrees. The composition of these objects is very complex and diversified, and affords very rich ground for exploration with a large telescope. Nebulæ exist in profusion and in every variety, and are intermingled with star-clusters varying in condensation from the compact globular form to groups more loosely scattered, and such as we often find in the Milky Way. Nearly three hundred nebulæ and clusters are included in the major "cloud," while more than fifty others closely outlie its borders. In the minor about forty such objects have been discovered. It is very strange to find them collected together in this manner; for in other regions of the firmament they are usually found separated, and certain classes appear to have their own special zones or localities of distribution. Sir J. Herschel pointed out that " globular clusters (except in one

* Humboldt says this " name is evidently derived from the voyage of Magellan, although he was not the first who observed them."

region of small extent) and nebulæ of regular elliptic forms
are comparatively rare in the Milky Way, and are found congregated in the greatest abundance in a part of the heavens
most remote possible from that circle, whereas in the Nubeculæ
they are indiscriminately mixed with the general starry ground
and with irregular though small nebulæ."

Double Nebulæ.—Instances are not wanting of conspicuous
double nebulæ. M. 51 and 76, near η Ursæ and θ Andromedæ,
may be classed in this category. There is a very interesting,
though a smaller object just W. of α and β Geminorum, or in
R.A. 7^h $18^{m.}6$, Dec. $+29^\circ$ $43'$. Two bright, round nebulæ
are separated by an interval of $28''$. These double nebulæ
are usually round, and are sometimes resolvable into stars.
Whether they are physical or mere optical pairs has yet to
be ascertained. So many examples exist that it seems highly
probable they have a real connection, though no motion has
yet been certainly detected to prove they are binary systems.
Such motion may, however, be very slow, and require observations extending over a much longer interval before it is
revealed.

Real Dimensions of Nebulæ and Clusters.—It may be readily
imagined that these objects are of immense size; for though
placed at distances of the utmost remoteness, they spread over
perceptible and comparatively large areas. Gore remarks
that, on the assumption that the globular cluster in Hercules
(M. 13) is $5'$ in diameter, and its parallax one tenth of a second,
its real diameter must be 3000 times the Sun's mean distance
from the Earth, or nearly 280 billions of miles! He further
points out that, though this group contains as many as
14,000 stars, according to Sir W. Herschel, yet each component may be separated many millions of miles from the
others, owing to the vast dimensions of the group. Details
like these are of course only approximate, as the distance of a
nebula or star-cluster has not yet been definitely ascertained.
The great nebulæ of Orion and Andromeda must extend over
prodigious regions in distance-space; but to quote figures
seems useless, in consequence of our inability to form just
conceptions of such immensity.

Round Nebulæ and Clusters.— Resolvable nebulæ and

clusters are frequently circular in outline. The central condensation is an indication of their globular form, though not always so, for many of these objects become suddenly much brighter in the middle, and show an apparently stellar nucleus. The material or stars forming the object cannot therefore be equally distributed. Where it suddenly brightens there is a great condensation, and in some cases several of these are evident in the form of bright rings, intensifying as the nucleus is approached. This irregular aggregation denotes the operation of "a force of condensation directed from all parts towards the centre of such systems." In regard to planetary nebulæ, they cannot be globular or they would exhibit a brightness increasing from the margin to the centre. Their even luminosity throughout affords the evidence of a special structure. Sir J. Herschel thought the planetary nebula (M. 97) near β Ursæ Majoris must either be in the form of a hollow globe or a flat circular disk lying perpendicular to the line of vision.

Description of Nebulæ and Clusters of Stars.—The latter objects are included in this chapter for several reasons. In a small telescope nearly all such clusters exhibit the aspect of nebulæ, and they have been catalogued with them, though, as already explained, some great distinctions are to be drawn. To the naked eye the cluster Præsepe, in Cancer, is usually visible as a patch of nebulosity, though on a very clear, dark night stars may be glimpsed sparkling about the spot, and a very small glass will suffice to show it as a nest of stars. This object, and some others of a more difficult character (their component stars being smaller and more compressed), are tabulated (I.) at the end of this chapter. A summary (II.) of globular clusters is also given, together with a list (III.) of nebulæ, a few of which are resolvable into stars[*]. It must be understood that these selections, though comprising many notable objects, are by no means exhaustive, the intention being merely to indicate some typical examples of fine nebulæ and clusters and of peculiarities of form or appearance, such as planetary, annular, elliptical, and centrally condensed nebulæ and loose, compressed, and globular clusters. Some of these

[*] I have selected the various objects in these lists from the New General Catalogue.

objects deserve individual references, as they present interesting details to the telescopic observer and come within the reach of moderate appliances.

Great Nebula in Andromeda (M. 31). This object has often been mistaken for a comet, for it is readily perceptible to the eye on a moonless night. It is very large—4° by 2½°, according to Bond, with a 15-inch refractor. He discovered a pair of dark streaks in the brightest region of the nebula, and these may be well seen in a 10-inch reflector. It is really triple ; for about 25' S. of the nucleus there is a very bright, round, resolvable nebula, discovered by Le Gentil, and a third, observed by Caroline Herschel, lies rather further to the N.W. Photographs by Roberts show dark rings dividing the bright interior parts of the nebula from the outer, and imparting to it a decided spiral tendency. This nebula has hitherto resisted attempts to resolve it into stars, though many hundreds have been seen in the foreground. But its spectrum is continuous, so that its stellar character is to be inferred.

Great Nebula in Orion (M. 42). Visible to the naked eye just below a line connecting β and ζ Orionis, and involving θ Orionis. It exhibits an extremely complicated structure, and many of its irregular branches and condensations may be discerned in small instruments. Sir W. Herschel failed to resolve this object into stars with his 4-foot reflector; but Lord Rosse, in 1844, thought he had effected it with his 6-foot mirror, though the conclusion was premature. The spectroscopic researches of Huggins have shown this nebula to be composed of incandescent gases, so that the stars telescopically observed in it are probably in the foreground and entirely disconnected from the nebulous mass. Effective photographs have been taken of it by Draper, Common, and Roberts. It certainly forms one of the grandest objects in the heavens.

The Planetary* Nebula (M. 97). Discovered by Mechain in 1781. In small telescopes it looks like a rather faint, round

---

* These forms are more numerous than the annular nebulæ. They often exhibit a blue colour, and the spectroscope shows them to consist of gas.

mass of nebulosity, somewhat brighter in the middle than at the edges. In Lord Rosse's telescope it shows many details, including a spiral arrangement and two dark spots in the middle inclosing bright, eye-like condensations. The margin is fringed with protuberances, and from its peculiar aspect this object has been called the " Owl " Nebula. Diameter between 155″ and 160″. It may readily be picked up $2\frac{1}{4}°$ S.E. of β Ursæ Majoris. It yields a gaseous spectrum.

In Draco at R.A. 17^h 58^m 36^s, Dec. $+66°$ 38′ there is a pretty small, but exceedingly bright planetary nebula. With a low power it looks like a star out of focus, but a high power expands it into a well-defined planetary disk. As observed in Lord Rosse's 3-ft. reflector on Sept. 17, 1873, this nebula exhibited " a round, well-defined disk of a full blue colour, light very equable, diameter 22″·4, surrounded by an extremely faint nebulosity. This is an excellent object of its class.

Spiral Nebula (M. 51). Discovered by Messier on Oct. 13, 1773. It is situated in Canes Venatici, and 4° S.W. from η Ursæ Majoris. An ordinary instrument will reveal it as a double nebula, and the two parts will be seen to differ greatly in size. Messier gave the distance separating them as 4′ 35″. Sir J. Herschel drew this object as a bright, centrally condensed nebula, surrounded by a dark space and then by a luminous ring divided through nearly one half of its circumference. Closely outlying this he placed a bright round nebula. Lord Rosse's 6-foot showed something very different. In April 1845 its spiral character was discovered ; coils of nebulosity were observed tending in a spiral form towards the centre, and the outlying nebula was seen to be connected with it. Some striking drawings have been published of this object. Those by Sir J. Herschel and Lord Rosse differ essentially, and would scarcely be supposed to represent the same nebula ; but when we reflect that the instruments used were respectively of 18 inches and 72 inches aperture, the cause of the disparity becomes evident.

Another fine example of a spiral nebula is M. 99, in the northern wing of Virgo, and 8° E.·of β Leonis. This object was discovered by Mechain; its spiral form of structure was detected by Lord Rosse in 1848. Diameter $2\frac{1}{2}′$. Like

M. 51 it gives a continuous spectrum and is resolvable into stars.

The Crab Nebula in Taurus (M. 1). Discovered by Bevis in 1731, and situated 1° N.E. of ζ Tauri.   Its diameter is $5\frac{1}{2}'$ by

Fig. 64.

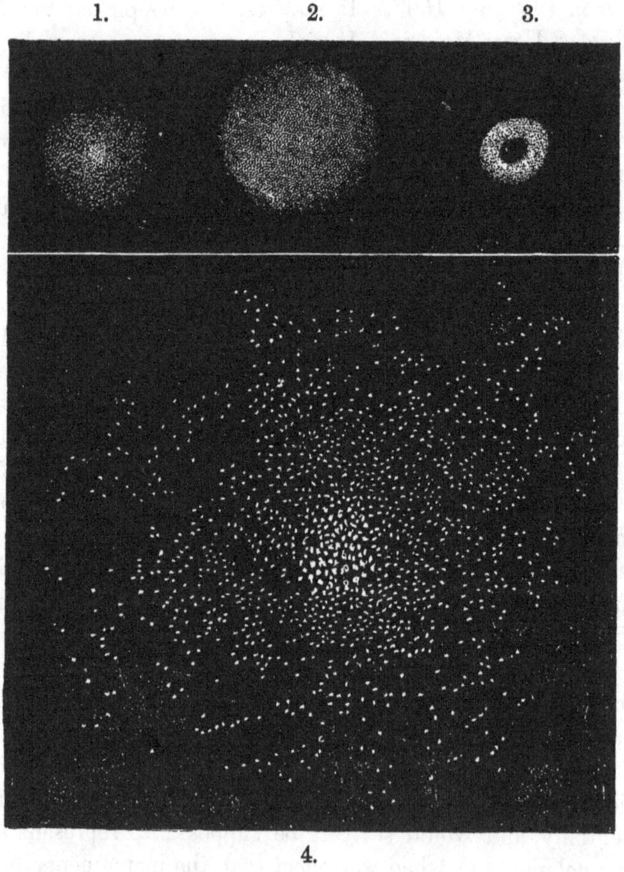

1.   2.   3.

4.

| 1. Nebula with bright centre. | 3. Ring-nebula in Lyra. |
| 2. Planetary Nebula. | 4. Star-cluster in Hercules. |

$3\frac{1}{2}'$.   An early drawing with Lord Rosse's telescope shows it with numerous radiations; whence it was termed the Crab Nebula, from the supposed resemblance : but later observations have given it quite another form.   In 1877 there was no trace of the nebulous arms : it appeared as a well-defined, oval nebula with some irregularities of structure. This object is very plain in small telescopes, and may be readily picked up from its proximity to ζ Tauri ; but in such instruments it is void of detail, and merely presents a pale, oval nebulosity. It has not been clearly resolved, though it has a mottled appearance, indicating a stellar composition, in large apertures.

The Dumb-bell Nebula (M. 27). Discovered by Messier in 1779, and situated in Vulpecula—a region very rich in small stars. Diameter about 7' or 8'. Its general form resembles a dumb-bell or hour-glass ; hence its name. Struve, Lord Rosse, and others have seen many stars in the nebulous mass, but the latter is not resolvable. I have seen seven stars in the nebula with a 10-inch reflector. Its peculiar shape is perceptible in a small instrument. This object frequently serves to illustrate books on Astronomy ; but the drawings by Sir J. Herschel, Lord Rosse, and others are curiously discordant, and show how greatly differences in telescopic power may affect the observed appearance of an object.

The Ring-Nebula in Lyra (M. 57). Discovered by Messier between the stars β and γ Lyræ. Diameter 80" by 60". This object is bright, though rather small, and it will stand high powers. The dark centre may possibly be glimpsed in a 3-inch refractor ; I have seen it readily in a $4\frac{1}{4}$-inch. It was at one period thought to be resolvable, but the spectroscope has negatived the idea, and shown it probably consists of nitrogen gas. A small star near the centre was frequently seen in Lord Rosse's telescope ; but the 36-inch refractor at Mount Hamilton reveals twelve stars projected on or within the ring, and several others have been suspected. There is a faint star exterior to the ring, and following it ; this is visible in small telescopes. The space within the ring is not quite dark, and the structure of the nebula is somewhat complicated as seen in large instruments. Another fine instance of an annular nebula may be found 3° preceding the 4th mag. star 41 Cygni, but it is not so large or conspicuous as that in Lyra. Its diameter is 47" by 41". Several stars were seen sparkling in it by Lord Rosse, who found the centre

was filled with faint light and the N. side of the ring broadest
and brightest.

Elliptical nebulæ are well represented by the pair (M. 81
and 82) about 2° E. of *d* (22) Ursæ Majoris. They are
separated by about 38' of declination, so that they may be
observed in the same field of a low-power eyepiece. The
preceding one is very bright and large (8' by 2'). The
following one is a ray or streak of nebulosity (7' by ¾'). On
May 21, 1871, the great Rosse telescope showed the latter as
a most extraordinary object, at least 10' in length and crossed
by several dark bands. Roberts photographed these nebulæ
on March 31, 1889. " The negative shows that the nucleus
[of M. 81], which has not a well-defined boundary, is surrounded by rings of nebulous or meteoric matter, and that
the outermost rings are discontinuous in the N.p. and S.f.
directions." M. 82 is " probably a nebula seen edgeways, with
several nuclei of a nebulous character involved, and the rifts
and attenuated places in it are the divisions of the rings that
would be visible as such if we could photograph the nebula
from the direction perpendicular to its plane " (' Monthly
Notices,' vol. xlix. p. 363). This fine pair may be easily
picked up in a small instrument. Another grand object
of this class (discovered by Caroline Herschel in 1783) lies in
R.A. 0^h $42^m·2$, Dec. $-25° 54'$, between the stars β Ceti and
α Sculptoris.

Globular clusters furnish us with many examples of easily
resolved and richly condensed balls of stars. M. 3 (discovered by Messier), M. 5 (discovered by G. Kirch), and
M. 13 (discovered by Halley) may be selected as amongst the
finest of these objects in the northern hemisphere. They are
severally visible to the naked eye, and may be found in a
telescope directed as follows :—M. 3, between Arcturus and
Cor. Caroli, and nearer the former; M. 5, 7° S.W. of α Serpentis and close to the double star 5 Serpentis ; M. 13, one
third the distance from η to ζ Herculis. They are brilliant
objects from 5' to 7' diameter. With power 60 on my 10-inch
reflector they are spangled with stars, though not fully resolved.
Smyth described M. 3 as consisting of about 1000 small stars,

blazing splendidly towards the centre. Webb hardly resolved it with a $3\frac{7}{10}$-inch refractor. Another fine object of this class (M. 80) will be encountered midway between α and β Scorpii. Sir W. Herschel described it as the richest and most compact group of stars in the sky, and it is noteworthy from the fact that a new star burst forth near its centre in 1860. There is a magnificent cluster, involving ω Centauri, which Sir J. Herschel considered as " beyond all comparison the richest and largest object of the kind in the heavens." It is visible to the naked eye as a 4th mag. star, but residents in northern latitudes are precluded from a view of it. Pegasus also supplies us with some fine clusters ; Maraldi picked up two in 1746 (M. 2 and 15), and these will respectively be found 5° N. of β and 4° W.N.W. of ϵ Pegasi. They are to be classed amongst the grandest objects of their kind.

In Cygnus, at R.A. 20^h 41^m 7^s, Dec. $+30°$ $19'$, near κ, and especially in the region immediately north-east, there exist irregular and extensive streams of faint nebulosity which may be said to form a telescopic milky way. Nebulæ and stars are curiously grouped together, forming a remarkable arrangement which will well repay study. To see these objects satisfactorily, a moonless night, free from haze or fog, should be chosen, and the power should be moderately low, or some of the more feeble nebulous films will be lost. The observer may spend some agreeable hours in sweeping over this region, which is one of the best in a wonderfully rich constellation.

Further Observations.—The fact that Swift has discovered many hundreds of nebulæ during the last few years affords indubitable proof that considerable numbers of these objects still await detection. No doubt they are generally small and faint, but it is necessary they should be observed and catalogued, so that our knowledge in this department may be rendered as complete as possible. New nebulæ are sometimes mistaken for expected comets, and occasionally give rise to misconceptions which would be altogether avoided were our data more exhaustive.

Those who sweep for nebulæ must have the means of deter-

z 2

mining positions, and a small telescope will be inadequate to
the work involved. A reflector of at least 10 inches, or
refractor of 8 inches, will be required; and a still larger
instrument is desirable, for to cope successfully with objects of
this faint character needs considerable grasp of light. The
power employed should be moderate; it must be high
enough to reveal a very small nebula, but not so high as to
obliterate a large, diffused, and faint nebula. In forming his
first catalogue of 1000 nebulæ, Sir W. Herschel used a Newtonian reflector of 18·7 inches aperture, power 157, field
15′ 14″; Swift's recent discoveries were effected with a
16-inch refractor and a periscopic positive eyepiece, power
132, field 33′. With a low power a very extensive field will
be obtained, and a large part of the sky may soon be examined,
but it will be done ineffectively. It is better to use a moderately high power, and thoroughly sweep a small region. The
work is somewhat different to comet-seeking; it must proceed
more slowly and requires greater caution, for every field has
to be attentively and steadily scanned. If the telescope is
kept in motion, a faint nebula will pass unseen. Some of
these objects are so feeble that they are only to be glimpsed
by averted vision. When the eye is directed, say, to the
E. side, a faint momentary glow comes from the W. side of
the field; but the observer discerns nothing on looking
directly for the object. On again diverting his gaze he
receives another impression of faint nebulosity from the same
point as before, and becomes conscious of its reality. Frequently, while comet-seeking, I meet with a small indefinite
object, the character of which cannot be determined by direct
scrutiny. On withdrawing the eye to another part of the
field, however, the mystery is solved. If the object is a
nebula, it flashes very distinctly on the retina; but if a
small cluster, the individual stars are seen sparkling in it.
These indirect views are usually so effective that the trouble
of applying higher powers is dispensed with.

The glow from a faint nebula or comet often apparently
fluctuates in a remarkable manner. Light-pulsations affect
it, causing the nebulosity to be intermittently visible. It

flashes out and enlarges, then becomes excessively feeble and indeterminate. The changes are not real, but due to the faint and delicate nature of the object, which is only fugitively glimpsed and presents itself differently with the slightest change in the manner of viewing it. Burnham has said there is no such thing as glimpsing an object ; but he is wrong. It is the intermediate step between steady visibility and absolute invisibility.

The work of sweeping for nebulæ is much delayed by the comparisons necessary for the identification of objects. The path may be smoothed by marking the known nebûlæ on a good chart, like Argelander's. The observer may then see, by reference, whether the objects he encounters have been picked up before. The labour of projecting all the nebulæ contained in the New General Catalogue would of course be considerable, and the observer will probably find it expedient to select certain regions for examination, and map such nebulæ as are included within their borders.

The discovery of new nebulæ offers an inviting field to amateurs. Vast numbers of these objects have escaped previous observation, for though the sky has been swept again and again, its stores have not been nearly exhausted. Mr. Barnard recently stated that with the powers of the great 36-inch refractor the number of known nebulæ (more than 8000) might readily be doubled ! As an example of their plentiful distribution in certain regions it may be mentioned that Mr. Burnham very recently discovered eighteen new nebulæ in a small area of 16′ by 5′·5 near the position in R.A. 13^h 38^m, Dec. 56° 20′ N. Near the pole of the northern heavens there exist many unrecorded nebulæ, as this region does not appear to have been thoroughly examined with a large instrument. It is often the case that several nebulæ are clustered near together. Whenever a new one is discovered the surrounding space should therefore be carefully surveyed in search of others. The region immediately outlying known objects may also be regarded as prolific ground for new discoveries. After several hours' employment in the work of searching for nebulæ or comets the eye is enabled to discern faint objects which were invisible at first, as it is in a better condition to

receive feeble impressions.   While comet-seeking in 1889 and
1890 I discovered ten new nebulæ, all near the N. pole, and
their approximate positions are given below :—

Ref. No.	Date of Discovery.	Position 1890. R.A.	Position 1890. Dec. +	Description.
1.	1889, Aug. 26	h m s 4 29 59	75° 25·2′	F. S., b. M., * 12, n. p.
2.	1890, Nov. 7	4 40 19	78 7·9	F., S., R.
3.	1890, Oct. 19	4 46 38	68 9·8	F., S., R., b. M.N., F. double * s. f.
4.	1890, Nov. 16	5 50 7	80 31·0	v. F., S.
5.	1890, Nov. 9	6 11 45	83 1·9	F., S., R., m. b. M.
6.	1890, Oct. 17	6 59 26	85 45·0	v.F., v. v. S., 12′ s. s.f. N.G.C. 2300.
7.	1890, Nov. 7	7 8 52	80 7·4	v. F., p. S., 22′ s. s. f. N.G.C. 2336.
8.	1890, Sept. 14	7 23 24	85 30·0	F., S., E., 46′ s. f. N.G.C. 2300.
9.	1890, Sept. 8	8 21 37	86 7·4	p. F., S., m. b. M., * n. f.
10.	1890, Aug. 23	8 34 30	85 54·4	F., S., R., g. b. M., near preceding.

Abbreviations:—F., faint; S., small; R., round; M., middle; N., nucleus,
E., extended; v., very; b., brighter; n., north; s., south; f., following;
p., pretty, preceding; m., much; g., gradually; *, star; N.G.C., New General
Catalogue.

No. 8 is placed centrally within a curious semicircle of
stars, thus :—

Fig. 65.

I.—CLUSTERS OF STARS.

No. N.G.C., 1888.	No. M., 1781.	Position, 1890.		Description.
		R.A.	Dec.	
		h m	° ′	
225.	...	0 37·1	+61 3	Stars 9th–10th mags. Between γ and κ Cassiopeiæ.
869.	...	2 11·3	+56 38	In Perseus. Stars 7th–14th mags.
1039.	34.	2 35·0	+42 18	A fine group, chiefly of 9th mag. stars.
1912.	38.	5 21·3	+35 44	Stars of various mags. In Auriga.
1960.	36.	5 29·0	+34 4	Stars of 9th–11th mags. Near 1912.
2099.	37.	5 45·1	+32 31	Stars and star-dust. 5° S. of θ Aurigæ.
2168.	35.	6 2·0	+24 21	Stars of 9th–16th mags. near η Geminorum.
2287.	14.	6 42·3	−20 38	Visible to naked eye. 4° S. of Sirius.
2437.	46.	7 36·8	−14 34	Nebula involved with cluster of 8th–13th mag. stars.
2477.	...	7 48·4	−38 16	Fine group of 12th mag. stars near ζ Argûs.
2516.	...	7 56·5	−60 34	Visible to naked eye. 200 stars of 7th–13th mags.
2547.	...	8 7·4	−48 56	Vis. n.e. Stars 7th–16th mags. Diameter 20′.
2548.	...	8 8·3	− 5 28	Stars of 9th–13th mags. In Monoceros.
2632.	44.	8 34·0	+20 22	Præsepe. Group of bright stars vis. n. e.
2682.	67.	8 45·2	+12 13	Large group of stars of 10th–15th mags.
4755.	...	12 47·1	−59 45	Very large group about κ Crucis.
6121.	4.	16 16·9	−26 16	Close to Antares. Group and line of stars through it.
6603.	24.	18 12·0	−18 28	Stars of 15th mag. 3° N. of μ Sagittarii.
6611.	16.	18 12·7	−13 50	Group of at least 100 stars of various mags.
6705.	11.	18 45·1	− 6 24	Stars of 11th mag. and fainter. Fine object.
6838.	71.	19 48·8	+18 29	Stars of 11th–16th mags. In Sagitta.
7243.	...	22 10·9	+49 20	A clustering of many bright stars.
7654.	52.	23 19·4	+61 0	Irrregular group of 9th–13th mag. stars.
7789.	...	23 51·5	+56 6	Grand cluster of 11th–18th mag. stars.

II.—Globular Clusters of Stars.

No. N.G.C., 1888.	No. M., 1781.	Position, 1890.		Description.
		R.A.	Dec.	
104.	...	h m 0 19·1	−72 42	Very large ; more than 15′ diameter.
288.	...	0 47·3	−27 11	Slightly elliptical. Stars 12th–16th mags.
362.	...	0 58·5	−71 26	Stars 13th–14th mags. Diameter 4′.
1261.	...	3 9·3	−55 38	Large. Stars and star-dust.
1851.	...	5 10·5	−40 10	Very bright and large. Fine object.
4147.	...	12 4·5	+19 9	Pretty large, round. Minute stars.
4590.	68.	12 33·7	−26 9	Much compressed group of 12th mag. stars.
5024.	53.	13 7·5	+18 45	Fine object. Chiefly 12th mag. stars.
5139.	...	13 20·2	−46 44	Very large; diameter 20′. At ω Centauri.
5272.	3.	13 37·1	+28 56	Visible to naked eye. Diameter 7′.
5634.	...	14 23·8	− 5 29	Very bright, considerably large. Round.
5904.	5.	15 13·0	+ 2 29	Visible naked eye. Stars 11th–15th mags. Diam. 5′.
5986.	...	15 38·8	−37 25	Stars of 13th–15th mags. In Lupus.
6093.	80.	16 10·5	−22 42	Stars of 14th mag. Between α and β Scorpii.
6205.	13.	16 37·7	+36 40	Visible naked eye. A grand object, in Hercules.
6218.	12.	16 41·5	− 1 45	Stars of 10th mag. and fainter. Diam. 4′.
6254.	10.	16 51·4	− 3 56	Stars of 10th–15th mags. Diameter 4′.
6266.	62.	16 54·2	−29 57	Stars of 14th–16th mags. In Scorpio.
6333.	9.	17 12·8	−18 24	Much compressed group of 14th mag. stars. Diam. 4′.
6341.	92.	17 13·8	+43 15	A mass of stars and star-dust. 7° N. π Herculis.
6402.	14.	17 31·8	− 3 11	Chiefly stars 15th mag. Diameter 4′.
6656.	22.	18 29·7	−24 0	Stars of 11th–15th mags. In Sagittarius.
6779.	56.	19 12·3	+30 0	Stars 11th–14th mags. Between β Cygni and γ Lyræ.
6809.	55.	19 33·0	−31 14	Fine, large, round cluster of stars 11th–13th mags.
7078.	15.	21 24·7	+11 41	Group of stars and star-dust. Diameter 5′.
7089.	2.	21 27·8	− 1 19	Exceedingly small stars. Diameter 5′.
7099.	30.	21 34·1	−23 41	Stars 12th–16th mags. Diameter 3′.

III.—NEBULÆ.

No. N.G.C., 1888.	No. M., 1781.	Position, 1890.		Description.
		R.A.	Dec.	
185.	...	h m 0 32·9	+47 44	Very large; pretty bright. Resolvable into stars.
224.	31.	0 36·7	+40 40	Great nebula in Andromeda.
253.	...	0 42·2	−25 54	Very, very large and bright. 24′ by 3′.
598.	33.	1 27·6	+29 57·1	Exceedingly bright and large. Nucleus.
650.	76.	1 35·4	+51 1	Very bright double nebula.
1365.	...	3 29·4	−36 30	Very bright and large. Elliptical.
1501.	...	3 57·5	+60 37	Pretty bright planetary nebula. Diam. 1′.
1514.	...	4 2·4	+30 29	Star of 9th mag. in nebula 3′ diameter.
1952.	1.	5 27·9	+21 56	Great Crab Nebula, near ζ Tauri. Stars.
1976.	42.	5 29·9	− 5 28	Great nebula involving θ Orionis.
1990.	...	5 30·6	− 1 16	Star (ε Orionis) involved in nebulosity.
2070.	...	5 39·5	−69 9	Visible to naked eye. Great "looped" nebula.
2392.	...	7 22·7	+21 8	Nebulous star of 9th mag.
2403.	...	7 26·2	+65 50	Very large and bright. Elliptical.
2655.	...	8 41·2	+78 38	Very bright. Condensed in the middle.
2681.	...	8 45·6	+51 44	Very large and bright. Centre=star 10th mag.
2683.	...	8 45·9	+33 51	Very large and bright. Elliptical.
2841.	...	9 14·4	+51 26	Very large and bright. Centre=star 10th mag.
2903.	...	9 25·9	+22 0	Large, elliptical, double nebula.
3031.	81.	9 46·5	+69 35	Exceedingly bright and large. Elliptical.
3034.	82.	9 46·7	+70 13	A bright ray. In field with preceding.
3242.	...	10 19·5	−18 5	Bright planetary nebula. Diameter 45″. Blue.
3372.	...	10 40·8	−59 6	Great nebula surrounding η Argûs.
3556.	...	11 5·4	+56 16	Large, rather bright. Elliptical.
3587.	97.	11 8·4	+55 37	Fine planetary nebula. Diameter 3′. Near β Ursæ Majoris.

III.—NEBULÆ (*continued*).

No. N.G.C., 1888.	No. M., 1781.	Position, 1890. R.A.	Dec.	Description.
3623.	65.	h m 11 13·2	+13 42	Large, bright, elliptical. Near following one.
3627.	66.	11 14·5	+13 36	Large elliptical nebula. Near β Leonis.
4254.	99.	12 13·3	+15 2	Very fine 3-branched spiral nebula.
4321.	100.	12 17·4	+16 26	Very large 2-branched spiral nebula.
4382.	85.	12 19·9	+18 48	Very bright; pretty large. Round.
4472.	49.	12 24·2	+ 8 37	Bright; round. Resolvable into stars.
4486.	87.	12 25·3	+13 0	Large; round. Bright centre. Third of three.
4565.	...	12 30·9	+26 36	A ray of bright nebulosity E. of Coma.
4736.	94.	12 45·7	+41 43	Large and bright. Nucleus. Resolvable.
5128.	...	13 19·0	−42 27	Very large and bright. Elliptical. Bifid.
5194.	51.	13 25·2	+47 46	Great spiral nebula near η Ursæ Maj.
5236.	83.	13 30·8	+29 18	Fine object. 3-branched spiral.
5367.	...	13 51·1	−39 27	Very large and bright. Condensed in the middle.
5907.	...	15 13·0	+56 44	Large, elliptical. Another very close to it.
6369.	...	17 22·6	−23 40	Pretty bright, small ring-nebula.
6514.	20.	17 55·7	−23 1	Bright; large. Trifid. Double star involved.
6523.	8.	17 56·9	−24 23	Bright, with loose cluster of stars.
6618.	17.	18 14·4	−16 13	Bright and extremely large. 2-hooked.
6720.	57.	18 49·5	+32 54	Ring-nebula between β and γ Lyræ.
6826.	...	19 41·8	+50 16	Pretty large and bright planetary nebula.
6853.	27.	19 54·9	+22 25	The "Dumb-bell" Nebula. Fine object.
6960.	...	20 41·1	+30 19	Large and bright. κ Cygni involved.
7009.	...	20 58·2	−11 48	Very bright, small, planetary nebula. Elliptical.
7662.	...	23 20·6	+41 56	Very bright, pretty small, planetary or ring-nebula.

NOTES AND ADDITIONS.

Large and Small Telescopes.

P. 19.—With reference to mountainous sites for large instruments, a remark in Sir Isaac Newton's ' Opticks' (1730) may be quoted :—" Telescopes . . . cannot be formed so as to take away that confusion of rays which arises from the tremors of the atmosphere. The only remedy is a most serene and quiet air, such as may perhaps be found on the tops of the highest mountains above the grosser clouds."

P. 27.—Lieut. Winterhalter, of the United States Navy, recently visited a large number of European observatories, and in describing that of Nice says :—" M. Perrotin declares that two hours' work with a large instrument is as fatiguing as eight with a small one, the labour involved increasing in proportion to the cube of the aperture, the chances of seeing decreasing in the same ratio, while it can hardly be said that the advantages increase in like proportion." The Nice Observatory, and its splendid instruments (including a 30-inch refractor), are due to the munificence of M. Bischoffsheim, who has expended about five million francs upon them.

P. 36.—The large refractor to be erected on Wilson's Peak of the Sierra Madre range of mountains, in Southern California, is to be 40 inches in diameter. The rough unground disks of glass are already in the hands of the Clarkes, of Cambridgeport, Mass. It is estimated that the complete object-glass and cell will cost something like $65,000, and the focal length of the instrument will be about 58 feet.

The Sun.

P. 100.—The last minimum of sun-spot frequency appears to have occurred at the middle of 1889. Conspicuous spots were very rare in the first half of 1890, but some fine groups were presented in the last half of the year. On Aug. 31 I saw a group extending over 113,000 miles in length, and on Nov. 27 there was another, which measured 123,700 miles.

P. 111.—Thompson's cardboard disks have been favourably spoken of as enabling observers to determine the positions of spots at any season of the year.

MERCURY.

P. 137.—At the meeting of the British Astronomical Association on Nov. 26, 1890, Mr. G. F. Chambers expressed his firm belief in the existence of an intra-Mercurial planet. The President (Capt. W. Noble) in his inaugural address pointed out the desirability of effecting further observations, both of Mercury and Venus, with a view to redetermine their rotation-periods. He justly remarked that moderately small instruments might be fittingly employed in the work, and that Schiaparelli's deductions (mentioned on pp. 142 and 149) ought to be accepted with extreme reserve pending their verification.

MARS.

P. 160.—Prof. W. H. Pickering observed some of the canals on Mars in 1890 with a 12-inch refractor, but was not able to double any of them. He says that, in examining these objects, the power employed should not " exceed one or two hundred." This is quite contrary to the advice of others, who recommend high magnifiers; and perhaps it accounts for Prof. Pickering's failure in recognizing the duple canals.

With the great 36-inch refractor Mr. Keeler saw, on July 5 and 6, 1890, some curious white spots on the edges of the gibbous limb of Mars, something similar to those visible on the unilluminated part of the lunar disk. The canals were observed as feeble diffused bands. The two satellites were seen by a lady visitor, though previously unaware of their existence.

P. 161.—The method of deriving the rotation-period of

Mars is exemplified by Mr. Proctor in the ' Monthly Notices,' vol. xxviii. p. 38. An interesting paper, " On the Determination of the Rotation-Period of Jupiter in 1835," will be found in the ' Memoirs,' vol. ix.

PLANETOIDS.

P. 167.—The 308th planetoid was discovered by Charlois on March 5, 1891.

JUPITER.

P. 170.—Dupret, in Algiers, saw Jupiter with the naked eye on Sept. 26, 1890, and following days, twenty minutes before sunset.

P. 191.—M. Guillaume, during a recent transit of the shadow of Jupiter's second satellite, observed a duplicate shadow, fainter than the ordinary one, which partly covered its southern side.

COMETS.

P. 250.—On Nov. 16, 1890, Dr. Spitaler, while looking for Zona's Comet with the 27-inch refractor of the Vienna Observatory, discovered a new and very faint comet only 23′ distant from the object of his search. That two of these bodies should be found almost simultaneously and so near together must be regarded as a very singular coincidence.

METEORS.

P. 261.—Mr. Proctor held the view that certain meteorites may have originally been ejected from the Sun. A recent writer thus summarizes our knowledge of them :—" That they are independent bodies, moving in orbits of their own in space ; that these dark bodies are abundant in the interplanetary spaces ; that those within the near range of solar or planetary attraction move with great velocity ; that many swarms of them follow well-known orbits ; and that, in general, their origin is undoubtedly the same as that of other celestial bodies " (' Sidereal Messenger,' June 1890, p. 284).

P. 267.—On May 2, 1890, a brilliant fireball, leaving a long train of fire and smoke, and exploding with a noise like thunder, was seen at many places in Northern Iowa, Minnesota, U.S.A. Some fragments of the meteor fell on a farm a few miles from the south line of Minnesota. The largest piece was sold by auction for $100, but it soon transpired that the person who sold it was only the lessee and not the owner of the ground on which the meteor fell. The aerial visitor and its purchase-money were therefore peremptorily seized by legal authorities, pending the decision of a Court of Justice as to the rightful ownership.

P. 267.—On December 14, 1890, at $9^h 42^m$ a large fireball of dazzling lustre, and giving a report like thunder, was widely observed in the southern parts of England. At the end-point the fireball appears to have been only 8 miles in height, and over a point near Brentwood, in Essex.

THE STARS.

P. 309.—Prof. Chandler, of Cambridge, Mass., estimates that the total number of variable stars visible with a common field-glass is about 2000, but with a large telescope there are probably hundreds of thousands within reach. He further states that quite five sixths of the variable stars are reddish in colour, and that the redness is usually a function of the length of the period of variation. The redder the star the longer its period.

P. 312.—In a recent communication to the Academy of Sciences, M. Lescarbault (the alleged discoverer of Vulcan in 1859) announced that on the night of January 11, 1891, he discovered a bright body in Leo which he could not identify in any star-map, and hence concluded it to be a new star, or one suddenly increased in brilliancy. The " new star," however, subsequently turned out to be the planet Saturn ! This ridiculous mistake (so easily avoidable with a little care) will naturally divest the supposed discovery of Vulcan of the importance attached to it by some writers, for M. Lescarbault obviously lacks the experience and caution necessary to command credit.

NEBULÆ AND CLUSTERS OF STARS.

P. 327.—Mr. Roberts, from a comparison of his photographs, has found distinct evidence of variability in the nucleus of the great nebula in Andromeda. In some of the photographs the nucleus is shown to be stellar, while in others there is no trace of this. Mr. Roberts remarks :—" We may reasonably infer that the nucleus of the nebula is variable, and that it will be practicable to study the character of the variability without the necessity of giving long exposures of the plates." The period of the variation has now to be determined, and it is advisable that telescopic observations of the nucleus should be made with the view of confirming the photographic results. It would be premature to regard the changes as demonstrated before they have been submitted to thorough investigation.

P. 327.—In the *Comptes Rendus* for March 2, 1891, M. Bigourdan has a paper on the variability of the nebula N.G.C. 1186, situated near Algol. This nebula was discovered by Sir W. Herschel in 1785, and though Sir J. Herschel re-observed it in 1831, Lord Rosse looked for it without success in 1854 and 1864. On Nov. 8, 1863, D'Arrest failed to detect the nebula, though he searched for it with assiduity at a time when the sky was very favourable. He was led to conclude that the object did not exist. M. Bigourdan finds that the nebula is again visible in the position indicated by the two Herschels, viz. R.A. 2^h 54^m 20^s, Dec. $+42°$ $10'$, he having observed it on Jan. 31 and Feb. 26, 1891. It is difficult to believe that this object could have escaped the scrutiny of Lord Rosse and D'Arrest in 1854, 1863, and 1864 ; hence the variation is probably real. The nebula may be easily found, as it is very near the binary B.D. $+42°$ (1123 G.C.), the position of which for 1891 is R.A. 2^h 58^m 6^s, Dec. $+42°$ $29'$ (' Nature,' March 12, 1891).

P. 329.—While examining the Pleiades on the night of November 14, 1890, Mr. Barnard discovered a new and considerably bright, round, cometary nebula $36''$ S. and $9''$ following Merope. The reason why this nebula has not been detected by photography is because it is so near Merope that the over-exposed light from the star obliterates it.　But it is certainly very strange that the object alluded to has never been telescopically discovered before ; for the Pleiades have been scrutinized repeatedly with all sorts of telescopes, and particularly since Tempel announced his discovery of a large faint nebula involving Merope in 1859.　Mr. Barnard says the new nebula is $30''$ in diameter, and that it is visible in a 12-inch refractor when Merope is hidden with a wire.

INDEX.

Printed in the United States
By Bookmasters